Mathematicians on Creativity

Mathematicians on Creativity

Edited by

Peter Borwein
Peter Liljedahl
Helen Zhai

Published and Distributed by
The Mathematical Association of America

Library of Congress Catalog Card Number 2014933023
Print edition ISBN: 978-0-88385-574-4
Printed in the United States of America
Current Printing (last digit):
10 9 8 7 6 5 4 3 2 1

SPECTRUM SERIES

The Spectrum Series of the Mathematical Association of America was so named to reflect its purpose: to publish a broad range of books including biographies, accessible expositions of old or new mathematical ideas, reprints and revisions of excellent out-of-print books, popular works, and other monographs of high interest that will appeal to a broad range of readers, including students and teachers of mathematics, mathematical amateurs, and researchers.

777 Mathematical Conversation Starters, by John de Pillis

99 Points of Intersection: Examples—Pictures—Proofs, by Hans Walser. Translated from the original German by Peter Hilton and Jean Pedersen

Aha Gotcha and Aha Insight, by Martin Gardner

All the Math That's Fit to Print, by Keith Devlin

Beautiful Mathematics, by Martin Erickson

Calculus and Its Origins, by David Perkins

Calculus Gems: Brief Lives and Memorable Mathematics, by George F. Simmons

Carl Friedrich Gauss: Titan of Science, by G. Waldo Dunnington, with additional material by Jeremy Gray and Fritz-Egbert Dohse

The Changing Space of Geometry, edited by Chris Pritchard

Circles: A Mathematical View, by Dan Pedoe

Complex Numbers and Geometry, by Liang-shin Hahn

Cryptology, by Albrecht Beutelspacher

The Early Mathematics of Leonhard Euler, by C. Edward Sandifer

The Edge of the Universe: Celebrating 10 Years of Math Horizons, edited by Deanna Haunsperger and Stephen Kennedy

Euler and Modern Science, edited by N. N. Bogolyubov, G. K. Mikhailov, and A. P. Yushkevich. Translated from Russian by Robert Burns.

Euler at 300: An Appreciation, edited by Robert E. Bradley, Lawrence A. D'Antonio, and C. Edward Sandifer

Expeditions in Mathematics, edited by Tatiana Shubin, David F. Hayes, and Gerald L. Alexanderson

Five Hundred Mathematical Challenges, by Edward J. Barbeau, Murray S. Klamkin, and William O. J. Moser

The Genius of Euler: Reflections on his Life and Work, edited by William Dunham

The Golden Section, by Hans Walser. Translated from the original German by Peter Hilton, with the assistance of Jean Pedersen.

The Harmony of the World: 75 Years of Mathematics Magazine, edited by Gerald L. Alexanderson with the assistance of Peter Ross

A Historian Looks Back: The Calculus as Algebra and Selected Writings, by Judith Grabiner

History of Mathematics: Highways and Byways, by Amy Dahan-Dalmédico and Jeanne Peiffer, translated by Sanford Segal

How Euler Did It, by C. Edward Sandifer

In the Dark on the Sunny Side: A Memoir of an Out-of-Sight Mathematician, by Larry Baggett

Is Mathematics Inevitable? A Miscellany, edited by Underwood Dudley

I Want to Be a Mathematician, by Paul R. Halmos

Journey into Geometries, by Marta Sved

JULIA: a life in mathematics, by Constance Reid

The Lighter Side of Mathematics: Proceedings of the Eugène Strens Memorial Conference on Recreational Mathematics & Its History, edited by Richard K. Guy and Robert E. Woodrow

Lure of the Integers, by Joe Roberts

Magic Numbers of the Professor, by Owen O'Shea and Underwood Dudley

Magic Tricks, Card Shuffling, and Dynamic Computer Memories: The Mathematics of the Perfect Shuffle, by S. Brent Morris

Martin Gardner's Mathematical Games: The entire collection of his Scientific American columns

The Math Chat Book, by Frank Morgan

Mathematical Adventures for Students and Amateurs, edited by David Hayes and Tatiana Shubin. With the assistance of Gerald L. Alexanderson and Peter Ross

Mathematical Apocrypha, by Steven G. Krantz

Mathematical Apocrypha Redux, by Steven G. Krantz

Mathematical Carnival, by Martin Gardner

Mathematical Circles Vol I: In Mathematical Circles Quadrants I, II, III, IV, by Howard W. Eves

Mathematical Circles Vol II: Mathematical Circles Revisited and Mathematical Circles Squared, by Howard W. Eves

Mathematical Circles Vol III: Mathematical Circles Adieu and Return to Mathematical Circles, by Howard W. Eves

Mathematical Circus, by Martin Gardner

Mathematical Cranks, by Underwood Dudley

Mathematical Evolutions, edited by Abe Shenitzer and John Stillwell

Mathematical Fallacies, Flaws, and Flimflam, by Edward J. Barbeau

Mathematical Magic Show, by Martin Gardner

Mathematical Reminiscences, by Howard Eves

Mathematical Treks: From Surreal Numbers to Magic Circles, by Ivars Peterson

A Mathematician Comes of Age, by Steven G. Krantz

Mathematicians on Creativity, edited by Peter Borwein, Peter Liljedahl and Helen Zhai

Mathematics: Queen and Servant of Science, by E.T. Bell

Mathematics in Historical Context, by Jeff Suzuki

Memorabilia Mathematica, by Robert Edouard Moritz

Musings of the Masters: An Anthology of Mathematical Reflections, edited by Raymond G. Ayoub

New Mathematical Diversions, by Martin Gardner

Non-Euclidean Geometry, by H. S. M. Coxeter

Numerical Methods That Work, by Forman Acton

Numerology or What Pythagoras Wrought, by Underwood Dudley

Out of the Mouths of Mathematicians, by Rosemary Schmalz

Penrose Tiles to Trapdoor Ciphers ... and the Return of Dr. Matrix, by Martin Gardner

Polyominoes, by George Martin

Power Play, by Edward J. Barbeau

Proof and Other Dilemmas: Mathematics and Philosophy, edited by Bonnie Gold and Roger Simons

The Random Walks of George Pólya, by Gerald L. Alexanderson

Remarkable Mathematicians, from Euler to von Neumann, by Ioan James

The Search for E.T. Bell, also known as John Taine, by Constance Reid

Shaping Space, edited by Marjorie Senechal and George Fleck

Sherlock Holmes in Babylon and Other Tales of Mathematical History, edited by Marlow Anderson, Victor Katz, and Robin Wilson

Six Sources of Collapse: A Mathematician's Perspective on How Things Can Fall Apart in the Blink of an Eye, by Charles R. Hadlock

Sophie's Diary, Second Edition, by Dora Musielak

Student Research Projects in Calculus, by Marcus Cohen, Arthur Knoebel, Edward D. Gaughan, Douglas S. Kurtz, and David Pengelley

Symmetry, by Hans Walser. Translated from the original German by Peter Hilton, with the assistance of Jean Pedersen.

The Trisectors, by Underwood Dudley

Twenty Years Before the Blackboard, by Michael Stueben with Diane Sandford

Who Gave You the Epsilon? and Other Tales of Mathematical History, edited by Marlow Anderson, Victor Katz, and Robin Wilson

The Words of Mathematics, by Steven Schwartzman

MAA Service Center
P.O. Box 91112
Washington, DC 20090-1112
800-331-1622 FAX 240-396-5647

Introduction

Why read this book?

Creativity and mathematics may be an oxymoron to the non-mathematical but it certainly isn't to productive mathematicians. Mathematical publications obscure mathematical creativity. Journal articles, books, and monographs do not present mathematicians in action, but show the work in finished form, with the footsteps of discovery (invention) carefully erased.

But mathematics rarely, if ever, emerges in finished form. The logicality of polished mathematics differs enormously from the more mysterious process of mathematical discovery. The more artful aspects of mathematics are elusive.

This book aims to shine a light on some of the issues of mathematical creativity. It is neither a philosophical treatise nor the presentation of experimental results, but a compilation of reflections from top-calibre working mathematicians. In their own words, they discuss the art and practice of their work. This approach highlights creative components of the field, illustrates the dramatic variation by individual, and hopes to express the vibrancy of creative minds at work.

No single recipe exists for mathematical discovery, but the collection included here suggests that some of the ingredients are hard work, experimentation, insight, tenacity, technical skill, mistakes, intuition, and good problems. The assembled quotations convey a taste of the creative side of mathematics. This book explores how people actively engaged in mathematics think about the undertaking as a creative and exciting pursuit.

The genesis of the project

In 1902, the first half of what eventually came to be a 30 question survey was published in the pages of *L'Enseignement Mathématique*. Édouard

Claparède and Théodore Flournoy, two French psychologists, who were deeply interested in the topic of mathematical creativity, authored the survey. Their hope was that a widespread appeal to mathematicians at large would incite enough responses for them to begin to formulate some conclusions about this topic. The first half of the survey centered on the reasons for becoming a mathematician (family history, educational influences, social environment, etc.), attitudes about everyday life, and hobbies. This was eventually followed up, in 1904, by the publication of the second half of the survey pertaining, in particular, to mental images during periods of creative work. The responses were sorted according to nationality and published in 1908, but for reasons that will soon become clear, quickly faded into obscurity.

By the time that Claparède and Flournoy's survey was published, one of the most noteworthy mathematicians of the time, Henri Poincaré, had already laid much of the groundwork for his own pursuit of this same topic. Consequently, he did not respond to the request published in *L'Enseignement Mathématique*, and for reasons that are not clear, neither did many of his peers. In fact, of those mathematicians who did respond to Claparède and Flournoy's survey, none could be called noteworthy ([49], 10). What was more damaging, however, was the fact that shortly after Claparède and Flournoy published their results, Poincaré gave a talk to the Psychological Society in Paris entitled *Mathematical Discovery*. At the time of the talk Poincaré stated that he was aware of Claparède and Flournoy's work, as well as their results, but stated that they would only confirm his own findings [124].

Another noted mathematician who did not respond to Claparède and Flournoy's survey was Poincaré's good friend Jacques Hadamard who would turn out to be the biggest critic of Claparède and Flournoy's work. Hadamard felt that the two psychologists had failed to adequately treat the topic of mathematical creation on two fronts; the first was the lack of comprehensive treatment of certain topics and the second was the lack of prominence on the part of the respondents. Further, Hadamard felt that as exhaustive as the survey appeared to be, it failed to ask some key questions—the most important of which was with regard to the reason for failures in the creation of mathematics. This seemingly innocuous oversight, however, led directly to what he termed "the most important criticism which can be formulated against such inquiries" ([49], 10). This leads to Hadamard's second, and perhaps more damning, criticism. He felt that only "first-rate men would dare to speak of"

([49], 10) such failures, and so, inspired by his good friend Poincaré's treatment of the subject Hadamard retooled the survey[1] and gave it to friends of his for consideration—mathematicians such as Henri Poincaré and Albert Einstein, whose prominence was beyond reproach. Ironically, the new survey did not contain any questions that explicitly dealt with failure. In 1943 he gave a series of lectures on mathematical invention at the École Libre des Hautes Études in New York City. These talks were subsequently published as *Essay on the Psychology of Invention in the Mathematical Field* [51].

Hadamard's classic work treats the subject of invention at the crossroads of mathematics and psychology. It provides not only an entertaining look at the eccentric nature of mathematicians and their rituals, but also outlines the beliefs of mid-twentieth-century mathematicians about the means by which they arrive at new mathematics. No such work representing the views of modern mathematicians exists. Thus, in 2002, 100 years after the work of Claparède and Flournoy began, we set out to resurrect a portion of Hadamard's classic survey with hopes of gaining insight from contemporary mathematicians about the creative process associated with doing mathematics.[2]

In order that he would be soliciting views from mathematicians whose creative processes are worthy of interest, Hadamard set excellence in the mathematical field as a criterion for participation in his study. Our interest, in keeping with Hadamard's standards, is in exceptional, rather than routine, creativity. To tease out inspiration and motivation for doing high level mathematics, we chose to survey the most prominent mathematicians: winners of the Fields medals, the Nevanlinna Prize, or the Ruth Lyttle Satter Prize in Mathematics. We have also polled members of The Royal Society, The American Academy of Arts and Sciences, and the Academié des Sciences with five questions drawn from Hadamard's survey. The particular questions we re-asked for which we were most interested in getting answers were:

1. Would you say that your principal discoveries have been the result of deliberate endeavor in a definite direction, or have they arisen, so to speak, spontaneously? Have you a specific anecdote of a moment of insight/inspiration/illumination that would demonstrate this? [Hadamard #9]

[1] See appendix A for a full version of Hadamard's survey.

[2] This project originated in some of the dissertation research of Peter Liljedahl in collaboration with Peter Borwein at Simon Fraser University.

2. How much of mathematical creation do you attribute to chance, insight, inspiration, or illumination? Have you come to rely on this in any way? [Hadamard #7]

3. Could you comment on the differences in the manner in which you work when you are trying to assimilate the result of others (learning mathematics) as compared to when you are indulging in personal research (creating mathematics)? [Hadamard #4, 15]

4. Have your methods of learning and creating mathematics changed since you were a student? How so? [Hadamard #16]

5. Among your greatest works, have you ever attempted to discern the origin of ideas that lead you to your discoveries? Could you comment on the creative process that led you to your discoveries? [Hadamard, #6]

The following questions were posed in a second round of questioning to elaborate on a recurring theme:

6. Your comments resonate with (or allude to) the idea of an AHA! or EUREKA! experience. I was wondering if you could comment a little bit more about such AHA! experiences? In particular, I would greatly appreciate it if you could answer the following question:

7. What qualities and elements of the AHA! experience serve to regulate the intensity of the experience? This is assuming that you have had more than one such experience and they have been of different intensities.

Not all respondents focused equally on these questions, of course. Some had more to say on certain points and nothing on others. One mathematician, Joseph Doob, claimed to be a bad subject, saying "Creative process is for the birds. I sat around wondering about what I was interested in. I am not a cosmic thinker." Many of his colleagues similarly commented about the superfluity of analysis at the moment of discovery—the mathematics itself is too captivating to be self-conscious about the creative process. Comments like this, in addition to somewhat freestanding essays, and reflections over lifelong mathematics careers, provided a rich picture of mathematical life and work, well beyond the intended subject of the so-called "AHA! moment." Naturally, schedule demands overwhelmed several mathematicians who declined to participate. This underscores the fact that today's academic world—where teaching loads, committee work, paperwork, and meetings make grant support a requirement for finding time to think—provides the setting for most creative mathematical work. But we will more assume this institutional setting than explore it.

From the varied and sometimes contradictory answers to our questions, a picture emerged indicating both the diversity of productive mathematical life and the intense curiosity of mathematicians about the workings of their own discipline. We were encouraged by their interest in the outcome of our project. Starting from this body of insightful first-hand ruminations, the project grew. We added a small selection of historical quotations related to issues of mathematical creativity. Mathematics, as we know it today, represents centuries of attempts, of failures, and of developments brought about by individuals working to understand and to possess something for themselves. Historical figures could not reply to the questions above, but their commentary lent a sense of depth, context, and background to the contemporary responses, while illustrating the scope of mathematical creativity over time. Artwork, manuscripts, and essays then illustrated various aspects of mathematical creativity.

How to read this book?

This book is probably best read by browsing. It is a gestalt rather than a whole. We wrestled with how to order the material and in the end decided to do it alphabetically. The sample of mathematicians is the tail end of a distribution. They are distinguished by their eminence and impact but do not lend themselves to systematic study. Nonetheless a significant sampling will give a good flavor of the subject. The quotations that we collected are marked with the symbol [ν], while the historical quotations are referenced accordingly. We do not draw conclusions but encourage readers to draw their own.

Acknowledgements

First, we would like to acknowledge the contributions of the mathematicians who so graciously responded to our queries. It had not been our intention to write a book when we began the project whose genesis is described in the introduction. We were surprised both by the thoughtfulness and quality of the discourse — particularly from such busy eminent mathematicians.

We would also like to acknowledge those mathematicians, living and dead, whose words on creativity in mathematics were culled from the literature. The amalgamation of their voices would not be possible had they not taken on the chore of articulating for the world their thoughts about this most elusive of phenomenon. Regardless of the nature of contribution, most of these aforementioned mathematicians are listed in Appendix B.

We would also like to thank Dr. Tom Archibald and Dr. Deborah Kent for their insightful historical guidance concerning creativity.

Contents

Alexander Craig Aitken (1895–1967)

I believe we are surrounded the whole time by marvellous powers, are immersed in them, closer than breathing, and I think that all great music, poetry, mathematics and real religion come from a world not distant but right in the midst of everything, permeating it. When I wish to do a feat of memory or calculation, or, sometimes, new mathematical discovery, I let slip some sort of cog and lie back in this world I speak of, not concentrating, but waiting in complete confidence for the thing desired to flow in. ([44], 19)

Jean d'Alembert (1717–1783)

Thus of all the sciences that pertain to reason, Metaphysics and Geometry are those in which imagination plays the greatest part. I ask pardon of those superior wits who are detractors of Geometry; doubtless they do not think themselves so close to it, although all that separates them perhaps is Metaphysics. Imagination acts no less in a geometer who creates than in a poet who invents... Of all the great men of antiquity, Archimedes is perhaps the one who most deserves to be placed beside Homer. ([7], 47–48)

George Andrews (1938–)

On the wall of my study is a poster picturing Stephen Leacock (the Canadian humorist) with the caption: ``I'm a great believer in luck! I find the harder I work the more I have of it." I would attribute my discovery of *Ramanujan's Lost Notebook* to luck arising out of hard work. [ν]

If you really want to learn something so that it is yours, so to speak, then the moments of insight come along with the hard slogging through computation and reflection. I should note that I do not feel that I create mathematics; I only discover it. So learning is discovery with more signposts provided by someone who has been down this path before. [ν]

There is a turn on the road along my drive home where I recall realizing instantly how the proof of a particularly troubling theorem had to go. [ν]

Don't force your preconceptions onto your research. Listen to what the mathematics is trying to tell you. Sometimes it speaks softly. [ν]

Computers are pencils with power steering. [ν]

... I really do not have any clear understanding of the creative process. I am just grateful that it happens! [ν]

Aristotle (384–322 BC)

The mathematical sciences particularly exhibit order, symmetry, and limitation; and these are the greatest forms of the beautiful. ([9], 1078 b)

Richard Askey (1933–)

Certain gaps in knowledge needed to be filled and my main role was to feel that these gaps could be filled. [ν]

It requires persistence and ability of high orders which are rare separately. [ν]

Behind beautiful and seemingly important formulas there must be deeper ideas. [ν]

Michael Atiyah (1929–)

A Eureka experience is characterized by suddenly realizing that you have found the missing piece of the jigsaw puzzle. Once found it is obviously right. The depth of the experience depends on how profound the ultimate result is. [ν]

My discoveries have all arisen indirectly, not by direction. They have come from asking myself questions about something that I think is mysterious or incompletely understood, and trying to get to the bottom of it. [ν]

Chance plays a role, but the key thing is to grab the chance. Here insight or intuition are very important. [ν]

Mathematics is an evolution from the human brain, creating the machinery with which it then attacks the outside world. It is our way of trying to reduce complexity into simplicity, beauty, and elegance. It is really very fundamental; simplicity is in the nature of scientific inquiry—we do not look for complicated things. ([133], 226)

I believe that if you do mathematics, you need a good relaxation that is not intellectual—being outside in the open air, climbing a mountain, working in your garden. But you actually do mathematics meanwhile. While you go for a long walk in the hills or you work in your garden, the ideas can still carry on. My wife complains, because when I walk she knows I am thinking of mathematics. ([133], 231)

Michael Atiyah

Wisława Szymborska, "Pi"

The admirable number pi:
three point one four one.
All the following digits are also initial,
five nine two because it never ends.
It can't be comprehended *six five three five* at a glance,
eight nine by calculation,
seven nine or imagination,
not even *three two three eight* by wit, that is, by comparison
four six to anything else
two six four three in the world.
The longest snake on earth calls it quits at about forty feet.
Likewise, snakes of myth and legend, though they may hold out a bit longer.
The pageant of digits comprising the number pi
doesn't stop at the page's edge.
It goes on across the table, through the air,
over a wall, a leaf, a bird's nest, clouds, straight into the sky,
through all the bottomless, bloated heavens.
Oh how brief—a mouse tail, a pigtail—is the tail of a comet!
How feeble the star's ray, bent by bumping up against space!
While here we have *two three fifteen three hundred nineteen*
my phone number your shirt size the year
nineteen hundred and seventy-three the sixth floor
the number of inhabitants sixty-five cents
hip measurement two fingers a charade, a code,
in which we find *hail to thee, blithe spirit, bird thou never wert*
alongside *ladies and gentlemen, no cause for alarm*,
as well as *heaven and earth shall pass away*,
but not the number pi, oh no, nothing doing,
it keeps right on with its rather remarkable *five*,
its uncommonly fine *eight*,
its far from final *seven*,
nudging, always nudging a sluggish eternity
to continue. ([154], 174)

Roger Bacon (1214–1292)

Mathematics is the gate and key of the sciences… Neglect of mathematics works injury to all knowledge, since he who is ignorant of it cannot know the other sciences or the things of this world. And what is worse, men who are thus ignorant are unable to perceive their own ignorance and so do not seek a remedy. [11]

Isaac Barrow (1630–1677)

… these Disciplines [mathematics] serve to inure and corroborate the Mind to a constant Diligence in Study; to undergo the Trouble of an attentive Meditation, and cheerfully contend with such Difficulties as lies in the Way. They wholly deliver us from a credulous Simplicity, most strongly fortify us against the Vanity of Scepticism, effectually restrain from a rash Presumption, most easily incline us to a due Assent, perfectly subject us to the Government of right Reason, and inspire us with Resolution to wrestle against the unjust Tyranny of false Prejudices. If the Fancy be unstable and fluctuating, it is to be poised by this Ballast, and steadied by this Anchor, luxuriant it is pared by this Knife; if headstrong it is restrained by this Bridle; and if dull it is roused by this Spur. The Steps are guided by no Lamp more clearly through the dark Mazes of Nature, by no Thread more surely through the intricate Labyrinths of Philosophy, nor lastly is the Bottom of Truth sounded more happily by any other Line. I will not mention how plentiful a Stock of Knowledge the Mind is furnished from these, with what wholesome Food it is nourished, and what sincere Pleasure it enjoys. But if I speak farther, I shall neither be the only Person, nor the first, who affirms it; that while the Mind is abstracted and elevated from sensible Matter, distinctly views pure Forms, conceives the Beauty of Ideas, and investigates the Harmony of Proportions; the Manners themselves are sensibly corrected and improved, the Affections composed and rectified, the Fancy calmed and settled, and the Understanding raised and excited to more divine Contemplation. All which I might defend by Authority, and confirm by the Suffrages of the greatest Philosophers. ([12], xxxi)

The *Mathematics*, I say, which effectually exercises, not vainly deludes nor vexatiously torments studious Minds with obscure Subtilties, perplexed Difficulties, or contentious Disquisitions; which overcomes without Opposition, triumphs without Pomp, compels without Force, and rules absolutely without the Loss of Liberty; which does not privately over-reach a weak Faith, but openly assaults an armed Reason, obtains a total Victory, and puts on inevitable Chains; whose Words are so many Oracles, and Works as many Miracles; which blabs out nothing rashly, nor designs anything from the Purpose, but plainly demonstrates and readily performs all Things within its Verge; which obtrudes no false Shadow of Science, but the very Science itself, the Mind firmly adheres to it, as soon as possessed of it, and can never after desert it of its own Accord, or be deprived of it by any Force of others: Lastly the Mathematics, which depend upon Principles clear to the Mind, and agreeable to Experience; which draws certain Conclusions, instructs by profitable Rules, unfolds pleasant Questions; and produces wonderful Effects; which is the fruitful Parent of, I had almost said all, Arts, the unshaken Foundation of Sciences, and the plentiful Fountain of Advantage to Human Affairs. ([12], xxviii)

Eric Temple Bell (1883– 1960)

This is precisely what common sense is for, to be jarred into uncommon sense. One of the chief services which mathematics has rendered the human race in the past century is to put 'common sense' where it belongs, on the topmost shelf next to the dusty canister labeled 'discarded nonsense.' ([14], 17–18)

Just as "beauty is its own excuse for being," so mathematics needs no apology for existing. ([16], 82)

Richard Bellman (1920–1984)

[Lefschetz and Einstein] had a running debate for many years. Lefschetz insisted that there was difficult mathematics. Einstein said that there was no difficult mathematics, only stupid mathematicians. I think that the history of mathematics is on the side of Einstein. ([18], 130)

The theory of elliptic functions is the fairyland of mathematics. The mathematician who once gazes upon this enchanting and wondrous domain crowded with the most beautiful relations and concepts is forever captivated.
([17], vii)

Elwyn Berlekamp (1940–)

It's somewhat like hunting or fishing. Luck plays a significant part, but your odds are much better if you have some judgement and experience about the regions where you think the big game are likely to be. [ν]

I had been working specifically on trying to prove that LR Hackenbush is NP-hard for several weeks, when most of the insight occurred to me while sitting in church during a long sermon to which I had tuned out. [ν]

I'm not very good at learning by reading. Most of my colleagues and students are far better at that than I am. Most of what I manage to assimilate I acquire by oral conversations and blackboard discussions, almost tutorials, from someone or other who knows it significantly better than I do. Usually I extract enough hints from him or her that I then manage to work it out and reconstruct some version of it myself. [ν]

You won't find anything if you're not prospecting, but if you keep your eyes open, you might stumble on gold when you thought you were looking for silver. [ν]

I think there is a widespread misperception that great discoveries are widely recognized as such when they are made. I think that's false. I've served on many prize and awards committees, and I've come to realize that most people grossly underestimate the importance of communicating their results. [ν]

In the commercial realm, [it's] called marketing, advertising, and/or sales... The similarities with the academic world are much greater than most academics are willing to admit. Sometimes the effective salesman gets as much or more credit for an idea than its inventor. [ν]

Lipman Bers (1914–1993)

I think that mathematics is very much like poetry. I think that what makes a good poem—a great poem—is that there is a large amount of thought expressed in very few words. In this sense formulas like

$$e^{\pi i} + 1 = 0$$

or

$$\int_{-\infty}^{\infty} e^{-x^2}\, dx = \sqrt{\pi}$$

are poems. ([5], 16)

Mathematics is an exceedingly cruel profession. You notice that if somebody has a bachelor's degree in chemistry, he describes himself as a chemist. But if somebody has been a professor of mathematics for ten years and you ask him, "Are you a mathematician?" he may say, "I'm trying to be one!" ...The standard is so high, and you never know whether you will be able to hack it. First you are afraid that you won't be able to understand your professors. Then you are afraid that you won't be able to write a thesis. When I went to Loewner to ask for a thesis topic, I expected him to grab me by the neck and say, "what makes you think you can write a thesis on mathematics? OUT!!!" ([5], 14)

I alternate between two attitudes [about attempts to communicate the beauty of mathematics to a wide audience]. Mondays, Wednesdays and Fridays I believe it can be done if we do it properly; Tuesdays, Thursdays and Saturdays I believe it cannot. ([5], 15)

I disagree with Adler, who wrote (in the *New Yorker*) that there is no point in being a mathematician unless you can be a great mathematician. That's nonsense. Mathematics is like a gothic cathedral. If you can build a little part of it, it is there—forever—in some sense. At least I have the illusion that it is so. ([5], 15)

Lipman Bers

What is the strength of mathematics? What makes mathematics possible? It is symbolic reasoning. It is like "canned thought." You have understood something once. You encode it, and then you go on using it without each time having to think about it. Now there may be people who are totally unable to follow symbolic reasoning—just as I am unable to carry a tune (and yet I do say to myself that I enjoy music). So you must try to explain mathematics without using any symbols. But this may be impossible. Without symbolic reasoning you cannot make a mathematical argument. ([5], 16)

A working mathematician is always a platonist. It doesn't matter what he says. He may not be a platonist at other times. But I think that in mathematics he always has that feeling of discovery. ([5], 19)

Usually people do their best work when they are young. And this is probably true of very good mathematicians. But in my case—and I think that most people who know my work will agree—what I did after forty was more interesting and more important than what I did before forty. ([5], 20)

George David Birkhoff (1884–1944)

... it is a faith in the uniformity of nature which remains the guiding star of the physicist just as for the mathematician it is a faith in the self-consistency of all mathematical abstractions, although these faiths are more sophisticated than ever before. The minds of both are tinged with an unwavering belief in the supreme importance of their own fields. The mathematician affirms with Descartes, *omnia apud me mathematica fiunt*—with me everything turns into mathematics; by this he means that all permanent forms of thought are mathematical. The physicist on his part is apt to think that there is no reality essentially other than physical reality, so that life itself is finally to be fully described in physical terms. ([19], 110)

David Blackwell

… [a mathematician] holds certain tacit beliefs and attitudes which scarcely ever find their way into the printed page… when he recalls that in the past the most difficult mathematical questions have been ultimately answered, he is inclined to believe with the great German mathematician, Hilbert, that every mathematical fact is provable. Besides all this, he attributes certain values to his results and their mathematical demonstrations; some theories seem important; some proofs are regarded as elegant, others as profound or original, etc. Such somewhat vague ideas illustrate what I would call mathematical faith. Nearly all the greatest mathematicians have been led to take points of view falling in this broad category, and have attached the deepest significance to them. ([19], 103)

David Harold Blackwell (1919–2010)

I'm interested in *understanding*, which is quite a different thing [from doing research]. And often to understand something you have to work it out yourself because no one else has done it. For example, I have gotten interested in Shannon's information theory. There are many questions that he left unanswered that were just crying out to be answered. The theory was incomplete so I worked on it with a couple of my colleagues because we wanted to know what happens in this case or that case. The drive was not to find something new. It would have been nicer if it had all been done. But since it hasn't been done, you just want to fill out the theory and make it complete. That's what I mean by being a dilettante. When I feel that my understanding of something has been rounded out pretty well, then I'm ready to move on to something else. ([4], 20)

Ralph Boas, Jr. (1912–1992)

Real mathematicians, except for a small number of geniuses, don't do anything *except* mathematics… Although I am fond of classical music, I never learned to play an instrument, and I am hopelessly unathletic. However, I grew up in the country and summered on Cape Cod, so I console myself by being able to do some things that my more cultivated colleagues probably can't. I do, for example, know how to sail a boat, shingle a roof, cut grass with a scythe, and fell a tree so that it will fall where I want it to. ([5], 30–31)

Some years ago, after I had given a talk, somebody said, "You seem to make mathematics sound like so much fun." I was inspired to reply, "If it isn't fun, why do it?" I am proud of the sentiment, even if it is overstated. ([5], 41)

A Brief Dictionary of Phrases Used in Mathematical Writing [119]

H. Pétard[1], Society of Useless Research

Since authors seldom, if ever, say what they mean; the following glossary is offered to neophytes in mathematical research to help them understand the language that surrounds the formulas. Since mathematical writing, like mathematics, involves many undefined concepts, it seems best to illustrate the usage by interpretation of examples rather than to attempt definition.

Analogue. This is an a. of: I have to have *some* excuse for publishing it.

Application. This is of interest in a.: I have to have *some* excuse for publishing it.

Complete. The proof is now c.: I can't finish it.

Details. I cannot follow the d. of X's proof: It's wrong. We omit the d.: I can't do it.

Difficult. This problem is d.: I don't know the answer. (Cf. Trivial.)

Generality. Without loss of g.: I have done an easy special case.

Ideas. To fix the i.: To consider the only case I can do.

Ingenious. X's proof is i.: I understand it.

Interest. It may be of i.: I have to have *some* excuse for publishing it.

Interesting. X's paper is i.: I don't understand it.

Known. This is a k. result but I reproduce the proof for the convenience of the reader: My paper isn't long enough.

Langage. Par abus de l.: In the terminology used by other authors. (Cf. Notation.)

Natural. It is n. to begin with the following considerations: We have to start somewhere.

New. This was proved by X but the following n. proof may present points of interest: I can't understand X.

Notation. To simplify the n.: It is too much trouble to change now.

Observed. It will be o. that: I hope you have not noticed that.

Obvious. It is o.: I can't prove it.

[1] H. Pétard is a pseudonym of Ralph Boas and Frank Smithies.

Reader. The details may be left to the r.: I can't do it.

Referee. I wish to thank the r. for his suggestions: I loused it up.

Straightforward. By a s. computation: I lost my notes.

Trivial. This problem is t.: I know the answer. (Cf. Difficult.)

Well-known. This result is w.: I can't find the reference.

Exercises for the student: Interpret the following.

1. I am indebted to Professor X for stimulating discussions.
2. However, as we have seen.
3. In general.
4. It is easily shown.
5. To be continued.

This article was prepared with the opposition of the National Silence Foundation.

Ralph Boas, Jr.

Maxime Bôcher (1867–1918)

I like to look at mathematics almost more as an art than as a science; for the activity of the mathematician, constantly creating as he is, guided though not controlled by the external world of the senses, bears a resemblance, not fanciful I believe but real, to the activity of an artist, of a painter let us say. Rigorous deductive reasoning on the part of the mathematician may be likened here to technical skill in drawing on the part of the painter. Just as no one can become a good painter without a certain amount of this skill, so no one can become a mathematician without the power to reason accurately up to a certain point. Yet these qualities, fundamental though they are, do not make a painter or mathematician worthy of the name, nor indeed are they the most important factors in the case. Other qualities of a far more subtle sort, chief among which in both cases is imagination, go to the making of a good artist or good mathematician. ([20], 133)

... there is what may perhaps be called the method of optimism which leads us either willfully or instinctively to shut our eyes to the possibility of evil. Thus the optimist who treats a problem in algebra or analytic geometry will say, if he stops to reflect on what he is doing: "I know that I have no right to divide by zero; but there are so many other values which the expression by which I am dividing might have that I will assume that the Evil One has not thrown a zero in my denominator this time." ([20], 134–135)

Salomon Bochner (1899–1982)

Mathematics is a form of poetry which transcends poetry in that it proclaims a truth; a form of reasoning which transcends reasoning in that it wants to bring about the truth it proclaims; a form of action, of ritual behavior, which does not find fulfilment in the act but must proclaim and elaborate a poetic form of truth. ([21], 191)

Enrico Bombieri (1940–)

Mathematics is the study of relations among objects. The nature of the objects is irrelevant; what matters is the internal structure of the relation, and there lies the universality of mathematics and the explanation why it is so important in all sciences. A mathematician is at the same time an explorer, an architect and an artist in this world of abstract ideas, moving freely around in his never ending quest for knowledge and perfection. His reward consists in sharing his knowledge and vision with others, and seeing his discoveries put to good use. [ν]

My approach to research consists in looking to the mathematical landscape, taking notice of the things I like and judge interesting and of those I don't care about, and then trying to imagine what should be next. If you see a bridge across a river, you try to imagine what lies on the other shore. If you see a mountain pass between two high mountains, you try to imagine what is in the valley you don't see yet but secretly know must be there. [ν]

I am not an architect or urban planner, rather more of a painter working on small paintings depicting what [where] the inspiration leads him. [ν]

Thus the first step of discovery consists for me in selecting an area of interest and good problems. How does one decide what is interesting? Usually, this is an instinctive process that takes very little time. [ν]

If I do something, I don't stop right away thinking that I have reached my goal. Rather, I stop and ask myself: What did I really find? What is next? Sometimes this is the first step for real progress. [ν]

My attitude towards mathematics is that most of it is lying out there, sometimes in hidden places, like gems encased in a rock. You don't see them on the surface, but you sense that they must be there and you try to imagine where they are hidden. Suddenly, they gleam brightly in your face and you don't know how you stumbled upon them. Maybe they always were in plain view, and we all are blind from time to time. [ν]

I don't see much difference between learning and creating mathematics, both steps are for me inextricably mixed. Reading a paper by another mathematician for me is comparable to a difficult hike in the mountains with the help of a guide, while in creating mathematics you are in a more familiar territory and a guide is not needed, you can follow your own path based on your experience and feeling. [ν]

I can compare the [Eureka] experience to putting together a very complicated puzzle without a blueprint, and suddenly you realize what it should be, and the pieces fall in the proper slot instantly. One does not need to put all the pieces in their proper places. Once you get the idea, the vision where exactly the bridge should be built, you know right away the litmus test to apply in order to confirm it. [ν]

That was the initial intuition, and in five minutes I knew it could be done and all the consequences it would entail. In conclusion, I think that for my best work I need intuition (or illumination, if it comes really suddenly) and also determination in reaching a goal … there have been occasions in which ideas came to me almost by chance or almost by themselves. For example, reading a paper one may see almost in a flash how to remove a stumbling block. [ν]

I worked three days and three nights never taking a rest save for eating a little and drinking coffee. [ν]

Howard Nemerov, "Figures of Thought"

To lay the logarithmic spiral on
Sea-shell and leaf alike, and see it fit,
To watch the same idea work itself out
In the fighter pilot's steepening, tightening turn
Onto his target, setting up the kill,
And in the flight of certain wall-eyed bugs
Who cannot see to fly straight into death
But have to cast their sidelong glance at it
And come but cranking to the candle's flame—

How secret that is, and how privileged
One feels to find the same necessity
Ciphered in forms diverse and otherwise
Without kinship—that is the beautiful
In Nature as in art, not obvious,
Not inaccessible, but just between.

It may diminish some our dry delight
To wonder if everything we are and do
Lies subject to some little law like that;
Hidden in nature, but not deeply so. ([105], 46)

Arthur Cayley (1821–1895)

It is difficult to give an idea of the vast extent of modern mathematics. The word 'extent' is not the right one: I mean extent crowded with beautiful detail—not an extent of mere uniformity such as an objectless plain, but of a tract of beautiful country seen at first in the distance, but which will bear to be rambled through and studied in every detail of hillside and valley, stream, rock, wood, and flower. But, as for everything else, so for a mathematical theory—beauty can be perceived but not explained. ([24], 496)

Arthur Cayley

Demetrios Christodoulou (1951–)

I do not attribute much to chance; I would attribute all to insight and illumination. In regard to illumination, I would like to add that in my case the best instances have been at night when I am lying in bed, somewhere between consciousness and sleep. It is during these times that I have the greatest power of concentration when all else except my subject lose reality. [ν]

I do not try to assimilate the results of others unless I am really interested in their work. In that case however, I will become completely absorbed, as if it is my own research. [ν]

I have always been attracted by the beauty of geometry and geometric thinking. [ν]

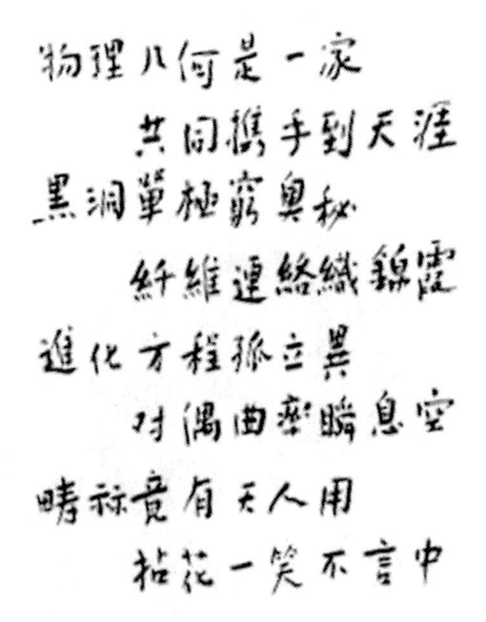

Figure 1. A poem written by Shiing Shen Chern. Translation by T. Y. Lam. on the occasion of the 2009 Chern Lectures at the University of California, Berkeley ([69], 865)

Physics and Geometry — to one family they belong,
Together, hand in hand, to the brink of the universe they've gone.
Black hole and monopole — deep mystery no longer shrouds,
Fibers and connections interlacing roseate clouds.
Evolution equations describe solitons,
Dual curvatures define instantons.
The art of Math subtly rules heaven and earth,
Fonding flowers with a smile, in quiet mirth.

Paul Cohen

Paul Cohen (1934–2007)

... once a problem is solved, I get a little bit bored. I guess that's the price you pay for being a problem solver. I am not really interested in problems that don't seem to stand out. ([5], 58)

I think to some extent that what I would like to do most is to take a problem that looks very complicated and find a solution that is ideal. I think that's what almost every mathematician really wants to do. And I would like to think that the solutions of the great problems in mathematics have that character. Hopefully the Riemann hypothesis, if it's solved, will have that character. ([5], 58)

I had, intuitively, a very strong philosophical feeling about the direction the proof should go; nevertheless, I felt totally frustrated. I was so low at one point that I stopped thinking about the problem for four or five months. ([5], 58)

John Horton Conway

John Conway (1937–)

I decided to take a crack at finding it since I knew a bit about groups. So I set up a schedule with my wife. We agreed that I would work on the problem on Wednesday nights from six until midnight and on Saturdays from noon until midnight. I started work on a Saturday, and made progress right away. At a half hour past midnight on that first Saturday, I came out with the problem solved! I had found the group! ([5], 8)

I suddenly realized that it was a good idea not to feel guilty; feeling guilty didn't do any good. Guilt just made it impossible to work.... So I decided for myself that from then on I wasn't going to work on something just because I felt guilty. If I was interested in some childish game, whereas previously I would have sort of looked around and wondered what my colleagues were thinking. ([5], 8)

... keep several things on the board, or at least on the back burner, at all times.... One of them is something where you can probably make progress.... If you work only on the really deep, interesting problems, then you're not likely to make much progress. So it's a good idea to have some less deep, less significant things, that nevertheless are not so shallow as to be insulting. ([5], 8)

Richard Courant (1888–1972)

It becomes the urgent duty of mathematicians, therefore, to meditate about the essence of mathematics, its motivations and goals and the ideas that must bind divergent interests together. ([26], 42)

Since the 17th century, physical intuition has served as a vital source for mathematical problems and methods. Recent trends and fashions have, however, weakened the connection between mathematics and physics; mathematicians, turning away from their roots of mathematics in intuition, have concentrated on refinement and emphasized the postulated side of mathematics, and at other times have overlooked the unity of their science with physics and other fields. In many cases, physicists have ceased to appreciate the attitudes of mathematicians. This rift is unquestionably a serious threat to science as a whole; the broad stream of scientific development may split into smaller and smaller rivulets and dry out. It seems therefore important to direct our efforts towards reuniting divergent trends by classifying the common features and interconnections of many distinct and diverse scientific facts. ([27], v–vi)

Richard Courant

Richard Courant

George Dantzig (1914–2005)

If I had known that the problems were not homework but were in fact two famous unsolved problems in statistics, I probably would not have thought positively, would have become discouraged, and would never have solved them. ([5], 68)

George Dantzig

The final test of a theory is its capacity to solve the problems which originated it. ([5], 71)

I didn't discover the linear programming model all in a flash. It evolved. ([5], 73)

I had no experience at the time with problems in higher dimensions, and I didn't trust my geometrical intuition. For example, my intuition told me that the procedure would require too many steps wandering from one adjacent vertex to the next. In practice, it takes few steps. In brief, one's intuition in higher dimensional space is not worth a damn! ([5], 77)

George Dantzig and Father

Carl De Boor (1937–)

I imagine that a mathematician's brain is similarly engaged in a ceaseless search for striking patterns in the ever-changing stream of ideas, making the mathematician aware of only the patterns found striking by the mathematician's pattern recognition process. [ν]

I have become more suspicious than I used to be about the originality of my ideas. [ν]

During the process 'insights' do appear seemingly spontaneously. However, this seems to me to be akin to artists looking at a landscape and being amazed by how interesting this or that view is—an amazement that ignores the fact that their artistic pattern recognition will only make them aware of certain views, namely those that are striking. [ν]

I learned from Fritz John and Heinz Kreiss (rather late in my mathematical life) that it is a waste of time to spend endless consecutive hours working on a problem, or filling page after page with calculations and scribbles. [ν]

When, after some considerable, quite nonproductive effort, usually while not at all consciously working on the problem, there appears, for no apparent reason, in your brain the answer to that problem. [ν]

Max Dehn (1878–1952)

At times the mathematician has the passion of a poet or a conqueror, the rigor of his arguments is that of a responsible statesman or, more simply, of a concerned father, and his tolerance and resignation are those of an old sage; he is revolutionary and conservative, sceptical and yet faithfully optimistic. These qualities do sometimes appear together in one person, but if you find them somewhat contradictory remember what C. F. Mayer makes Ulrich von Hutten say, and what the mathematician may claim for himself:

Ich bin kein ausgeklügelt Buch,
Ich bin ein Mensch mit seinem Widerpruch.

(I am not a contrived book, but a human being with all its contradictions). ([30], 26)

Pierre René Deligne (1944–)

For me, and in this I am inspired by Grothendieck, the ideal is a proof which is trivial, because it has been preceded by the "correct" definitions. [ν]

My main way of proceeding is trying to get an understanding, rather than trying to solve a problem. I first of all want to understand how things stand together, and what symmetries they have—while keeping in the back of my mind questions I care about, so as to be alert if in the whole panorama a way of attack presents itself. [ν]

Writing up is painful, but leads to understandings which can be reused. [ν]

A shorter analog: after a partial step, wondering: what does this argument really mean? [ν]

When reading, my first questions are: what are the tools used, do the definitions make sense, what are the data, what is only supposed to exist. What I try to remember is "picture, possibly conflicting, with caveat, to know what is true, as well as what is provable—and which tools can be used to prove such and such." [ν]

René Descartes (1596–1650)

If Fear of God is the beginning of wisdom. The actors, called to the scene, in order to hide their flaming cheeks, don a mask. Like them, when climbing on stage in the theatre of the world, where, thus far, I have only been a spectator, I advance masked. At the time of my youth, witnessing ingenious discoveries, I asked myself whether I could invent all on my own, without leaning on the work of others. Henceforth, little by little, I became aware that I was proceeding according to determined rules. Science is like a woman: if faithful, she stays by her husband, she is honored; if she gives herself to everyone, she is degraded. ([1], 1–2)

... the two operations of our understanding, intuition and deduction, on which alone we have said we must rely in the acquisition of knowledge. [32]

If I found any new truths in the sciences, I can say they follow from, or depend on, five or six principal problems which I succeeded in solving and which I regard as so many battles where the fortunes of war were on my side. ([31], 47)

René Descartes

Persi Diaconis (1945–)

Inventing a magic trick and inventing a theorem are very, very similar activities in the following sense. In both subjects you have a problem that you're trying to solve with constraints. In mathematics, it's the limitations of a reasoned argument with the tools you have available, and with magic it's to use our tools and sleight of hand to bring about a certain effect without the audience knowing what you're doing. The intellectual process of solving problems in the two areas is almost the same. When you're inventing a trick, it's always possible to have an elephant walk on stage, and while the elephant is in front of you, sneak something under your coat, but that's not a good trick. Similarly with mathematical proof, it is always possible to bring out the big guns, but then you lose elegance, or your conclusions aren't very different from your hypotheses, and it's not a very interesting theorem. ([4], 71)

Persi Diaconis

When I was young and doing magic, if I heard that an Eskimo had a new way of dealing a second card using snowshoes, I'd be off to Alaska. I spent ten years doing that, traveling around the world, chasing down the exclusive, interesting secrets of magic. An analog of that in my second career is not just doing any one thing. For example, my thesis was in number theory, and some might think that I do that kind of mathematics. I've done a fair amount of classical mathematical statistics, so you might think I do that. I have worked in philosophy of statistics, psychology of vision and pure group theory. What happens now is that if I hear about a beautiful problem, and if that means learning some beautiful math machine, then, boy, I'm off in a second to learn the secrets of the new machine. I'm just following the mathematical wind. ([4], 72)

Paul Dirac (1902–1984)

I think that there is a moral to this story, namely that it is more important to have beauty in one's equations than to have them fit experiment. If [Erwin] Schrödinger had been more confident of his work, he could have published it some months earlier, and he could have published a more accurate equation… It seems that if one is working from the point of view of getting beauty in one's equations, and if one has really a sound insight, one is on a sure line of progress. If there is not complete agreement between the results of one's work and experiment, one should not allow oneself to be too discouraged, because the discrepancy may well be due to minor features that are not properly taken into account and that will get cleared up with further developments of the theory. ([33], 47)

Lejeune Dirichlet (1805–1859)

In mathematics as in other fields, to find one self lost in wonder at some manifestation is frequently the half of a new discovery. ([34], 233)

David Donoho (1957–)

Obviously you work like hell and once in a while you notice something really unexpected. [v]

[Researchers] may become obsessed, reflect very deeply on it, etc. This is a prerequisite for being an outstanding researcher. Education has nothing to do with this. It has to do with *not* becoming deeply attached to some topic or topics but developing an all-purpose methodology that can be applied to topics across subject. And working diligently and in an organized way. [v]

[M]ost results are just about choosing the right topic and then working hard. Ahh! Who knows? I would not want to discount manic-depression and seasonal affective disorder as important biological issues. [ν]

First, you can only do something intellectually worthwhile by devoting an embarrassingly extreme amount of time preparing yourself both within a specialty and by reading voraciously and very broadly outside the specialty as well. If they only knew the amount of dedicated, concentrated work involved, most people would be shocked and repelled at the sacrifice involved. (Of course, a few people will have exactly the kind of obsessive personality that drives them to this kind of effort; the rest would find it an unimaginable deprivation) I think this is true even of the greatest mathematicians, and I'll bet it is true of greatness in many other fields as well. [ν]

You have scholars who spend 10 to 15 years investing every bit of their personalities in a single project. Real progress in research comes not from having many people know your work, but from having a few people understanding your work deeply. This is why small-scale meetings are so important. When it really works it's intense, something like the Vulcan mind-meld on 'Star Trek'. [147]

Joseph Doob (1910–2004)

I am afraid I am a bad subject for your investigation. I am not a cosmic thinker. [ν]

"Creative process" is for the birds. I just sat around and wondered about what I was interested in. [ν]

Freeman Dyson (1923–)

On being asked what he meant by the beauty of a mathematical theory of physics, Dirac replied that if the questioner was a mathematician then he did not need to be told, but were he not a mathematician then nothing would be able to convince him of it. ([13], 16)

Most of the mathematicians I knew were rather lonely people... Among the young mathematicians I found a high proportion were too crazy for my taste. There were a lot of them who were a bit crazy. ([5], 12)

The bottom line for mathematicians is that the architecture has to be right. In all the mathematics that I did, the essential point was to find the right architecture. It's like building a bridge. Once the main lines of the structure are right, then the details miraculously fit. The problem is the overall design. ([5], 20)

Freeman Dyson

Some mathematicians are birds, others are frogs. Birds fly high in the air and survey broad vistas of mathematics out to the far horizon. They delight in concepts that unify our thinking and bring together diverse problems from different parts of the landscape. Frogs live in the mud below and see only the flowers that grow nearby. They delight in the details of particular objects, and they solve problems one at a time... Mathematics needs both birds and frogs. Mathematics is rich and beautiful because birds give it broad visions and frogs give it intricate details. Mathematics is both great art and important science, because it combines generality of concepts with depth of structures. It is stupid to claim that birds are better than frogs because they see farther, or that frogs are better than birds because they see deeper. The world of mathematics is both broad and deep, and we need birds and frogs working together to explore it. ([38], 212)

Sir Arthur Eddington (1882–1944)

We have found a strange footprint on the shores of the unknown. We have devised profound theories, one after another, to account for its origin. At last, we have succeeded in reconstructing the creature that made the footprint. And lo! it is our own. ([28], 231)

Bradley Efron (1938–)

At first I'm terribly confused, but after awhile I chip away at my wrong ideas until I'm left with an answer. So I think I'm working in the sculptor mode, rather than the inspired painter. [ν]

I haven't been very interested in my own mental processes. [ν]

I'm afraid I ran out of Aha's a long time ago. They are dangerous after a certain age anyway, leading easily to a guru complex. [ν]

Inspiration starts things, but only hard work really gets anywhere. [ν]

Albert Einstein (1879–1955)

I imagine myself becoming a professor in those branches of the natural sciences, choosing the theoretical parts of them. Here are the reasons that have brought me to this plan. Above all, it is my disposition for abstract and mathematical thought, and my lack of imagination and practical ability. ([35], 13)

I collect nothing but unanswered correspondence and people who, with justice, are dissatisfied with me. But can it be otherwise with a man possessed? As in my youth, I sit here endlessly and think and calculate, hoping to unearth deep secrets. The so-called Great World, i.e., men's bustle, has less attraction than ever, so that each day I find myself becoming more of a hermit. ([35], 17)

As for the search for truth, I know from my own painful searching, with its many blind alleys, how hard it is to take a reliable step, be it ever so small, towards the understanding of that which is truly significant. ([35], 18)

For the creation of a theory the mere collection of recorded phenomena never suffices—there must always be added a free invention of the human mind that attacks the heart of the matter. ([35], 29)

Albert Einstein

Every one who is seriously involved in the pursuit of science becomes convinced that a spirit is manifest in the laws of the Universe—a spirit vastly superior to that of man, and one in the face of which we with our modest powers must feel humble. In this way the pursuit of science leads to a religious feeling of a special sort, which is indeed quite different from the religiosity of someone more naive. ([35], 33)

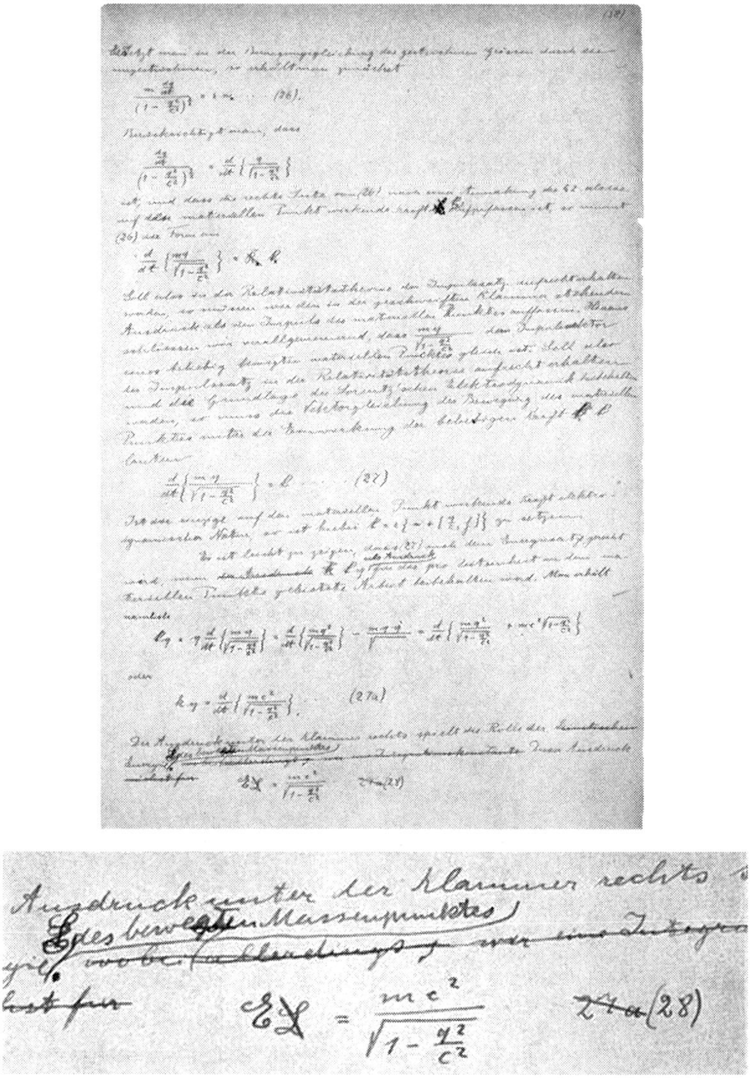

Figure 2. The world's most famous equation, from a manuscript on special relativity theory written by Albert Einstein in 1912. ([141], 103)

What Artistic and Scientific Experience Have in Common:
Where the world ceases to be the scene of our personal hopes and wishes, where we face it as free beings admiring, asking, and observing, there we enter the realm of Art and Science. If what is seen and experienced is portrayed in the language of logic, we are engaged in science. If it is communicated through forms whose connections are not accessible to the conscious mind but are recognized intuitively as meaningful, then we are engaged in art. Common to both is that loving devotion to that which transcends personal concerns and volition. ([35], 37)

Music does not *influence* research work, but both are nourished by the same source of longing, and they complement one another in the release they offer. ([35], 78)

Experience remains, of course, the sole criterion of the physical utility of a mathemaitcal construction. But the creative principle resides in mathematics. In a certain sense, therefore, I hold it true that pure thought can grasp reality, as the ancients dreamed. ([40], 398)

Erdős, who put in 19-hour days proving and conjecturing, denied that he fell asleep during mathematics conferences. "I wasn't sleeping," he would say. "I was thinking." ([64], 154)

Albert Einstein

Paul Erdős (1913–1996)

There's an old debate about whether you create mathematics or just discover it. In other words, are the truths already there, even if we don't yet know them? If you believe in God, the answer is obvious. Mathematical truths are there in the SF's mind, and you just rediscover them. Remember the limericks:

There was a young man who said, 'God,
It has always struck me as odd
That the sycamore tree
simply ceases to be
When there's no one about in the quad.'

Dear Sir, Your astonishment's odd;
I am always about in the quad:
And that's why the tree
Will continue to be,
Since observed by,
Yours faithfully, God.'

(The SF is the Supreme Fascist, the Number-One Guy Up There, God, who was always tormenting Erdős by hiding his glasses, stealing his Hungarian passport, or, worse yet, keeping to Himself the elegant solutions to all sorts of intriguing mathematical problems.) ([64], 26)

Paul Erdős

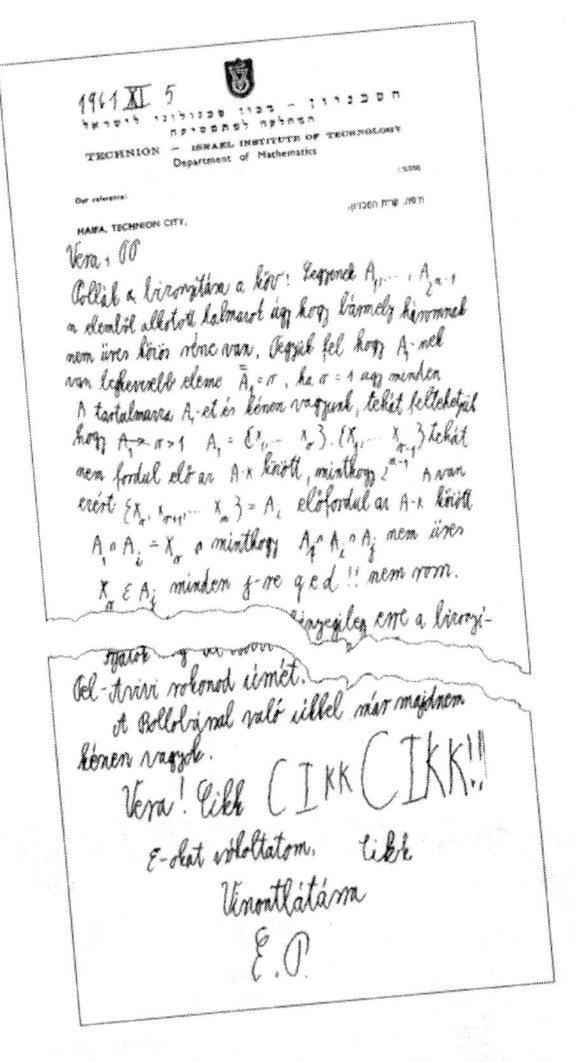
1961 XI 5

TECHNION — ISRAEL INSTITUTE OF TECHNOLOGY
Department of Mathematics

Our reference:

HAIFA, TECHNION CITY.

Vera! Cikk CIKK CIKK!!

E.P.

Figure 3. This sample letter was written in 1961 at the Technion (Israel). Addressed to Vera Sós and her husband, Paul Turán, the letter is written in Hungarian: Vera and TP. Polláḱs proof runs as follows: Let $A_1 \ldots$ *The second page describes a proof by Rado. In the concluding paragraphs Erdős asks Turán to provide the address of a relative, reports that an article with Bollobás is nearly done, and emphatically urges Vera to finish an article (CIKK = article). Closing: Give my love to the* ε *-s[a reference to the two children of Sós and Turán], Good-bye, E.P. ([10], 69)*

Leonhard Euler (1707–1783)

It will seem not a little paradoxical to ascribe a great importance to observations even in that part of the mathematical sciences which is usually called Pure Mathematics, since the current opinion is that observations are restricted to physical objects that make impression on the senses. As we must refer the numbers to the pure intellect alone, we can hardly understand how observations and quasi-experiments can be of use in investigating the nature of the numbers... The kind of knowledge which is supported only by observations and is not yet proved must be carefully distinguished from the truth; it is gained by induction, as we usually say. Yet we have seen cases in which mere induction led to error. Therefore, we should take great care not to accept as true such properties of the numbers which we have discovered by observation and which are supported by induction alone. Indeed, we should use such a discovery as an opportunity to investigate more exactly the properties discovered and to prove or disprove them; in both cases we may learn something useful. ([43], 459)

Till now the mathematicians tried in vain to this day to discover some order in the sequence of prime numbers and we have reason to believe that it is a mystery into which the human mind shall never penetrate. ([43], 241)

Leonhard Euler

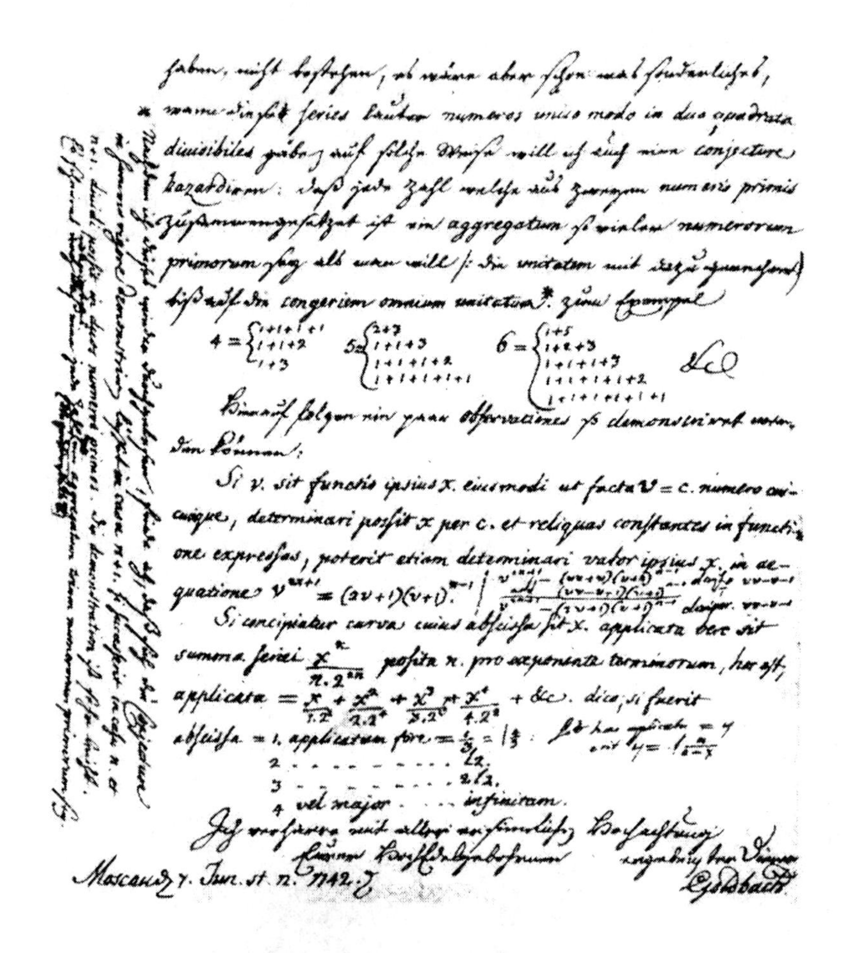

Figure 4. Christian Goldbach's letter to Leonhard Euler, dated June 7, 1942: in the margin, Goldbach conjectures that every number greater than two an be written as the sum of three primes (he considered one to be a prime).

Gerd Faltings (1954–)

One has to search the terrain until one finds an opening (or gives up), and where that is cannot be planned. [ν]

However, I do not think that this is quite by chance. Namely one has to spend much time on the subject before one gets inspiration. [ν]

Learning work of others of course means following their thoughts, as opposed to thinking for oneself. However I usually try whether I can find my own proofs for the assertions before I dig into their details. [ν]

...the feeling experienced when discovering a new insight... The intensity is of course regulated by the magnitude of the new insight, and/or the desperateness before. Also I should mention that these experiences are not so uncommon, but many of them do not last long because often the new insight later turns out to be false. [ν]

In some sense all insights come suddenly, usually in some impure form which is clarified later. [ν]

Solomon Feferman (1928–)

Understanding others is often a painful process until one suddenly goes beyond the details and sees whole what's going on. Many things are so difficult or so foreign that that never happens. Teaching other people's mathematics is the best way to achieve understanding. [ν]

Usually, specific endeavour in a definite direction, but often the difficulties met in the process are overcome only through insight/inspiration/illumination. But some lines of pursuit came about through the latter, with the sudden idea that it might be possible to do something of a certain kind and/or in a certain way. [ν]

Charles Fefferman (1949–)

It seems to me that insight is essential, and even the best mathematicians need luck, but if enough people do mathematics, luck will happen sooner or later. Of course, luck alone almost never succeeds. Hard work tilts the odds considerably. [ν]

When I read someone else's work, I first just read the statement of the theorem, and try to prove it myself. If I fail (I usually do), then I turn to the paper I'm reading for a hint. After reading a little, I try again, and the process repeats. Nevertheless, it seems to me utterly different from doing research, because in reading, one knows in advance that the desired result is true, and you are allowed to "give up" by reading the paper at hand. When doing research, you don't know whether whatever you are trying to prove is true, and, if you are 100% hopelessly stuck, you still have to keep trying. [ν]

I pick my research problems motivated only by my own tastes and interests. As a student, I was influenced by knowing that this or that was a famous problem. I don't know whether that's good or bad, but it's true. Otherwise, of course I am a bit slower and less energetic now than in my student days, but also a bit less ignorant. [ν]

I like to lie down on the sofa for hours at a stretch thinking intently about shapes, relationships and change—rarely about numbers as such. I explore idea after idea in my mind, discarding most. When a concept finally seems promising, I'm ready to try it out on paper. But first I get up and change the baby's diaper... New ideas are not easy to find. If you are lucky enough to be working on an idea which is actually right, it can take a long time before you know that it's right. Conversely, if you are going up a blind alley, it can also take a long time before you find out. You can end up saying 'Oops, I've been working for years on something wrong.' A good mathematician must have the courage to take a lot of work and throw it away. [110]

Wendell Fleming (1928–)

Both chance and insight are quite important. Chance will favour only those who are prepared. Also, one must expect to consider many ideas, which turn out later to be failures. [ν]

I should add that I still enjoy working on math at age 73. Currently I am collaborating with an economist on a problem of international debt and finance. [ν]

Joseph Fourier (1768–1830)

Profound study of nature is the most fertile source of mathematical discoveries. Not only has this study, in offering a determinate object to investigation, the advantage of excluding vague questions and calculations without issue; it is besides a sure method of forming analysis itself, and of discovering the elements which it concerns us to know, and which natural science ought always to preserve: these are the fundamental elements which are reproduced in all natural effects.

Considered from this point of view, mathematical analysis is as extensive as nature itself; it defines all perceptible relations, measures times, spaces, forces, temperatures; this difficult science is formed slowly, but it preserves every principle which it has once acquired; it grows and strengthens itself incessantly in the midst of the many variations and errors of the human mind. ([45], 7–8)

Its chief attribute is clearness; it has no marks to express confused notions. It brings together phenomena the most diverse, and discovers the hidden analogies which unite them. If matter escapes us, as that of air and light, by its extreme tenuity, if bodies are placed far from us in the immensity of space, if man wishes to know the aspect of the heavens at successive epochs separated by a great number of centuries, if the actions of gravity and of heat are exerted in the interior of the earth at depths which will be always inaccessible, mathematical analysis can yet lay hold of the laws of these phenomena. It makes them present and measurable, and seems to be a faculty of the human mind destined to supplement the shortness of life and the imperfection of the senses; and what is still more remarkable, it follows the same course in the study of all phenomena; it interprets them by the same language, as if to attest the unity and simplicity of the plan of the universe, and to make still more evident the unchangeable order which presides over all natural causes. ([45], 7–8)

Joseph Fourier

Galileo Galilei (1564–1642)

[The universe] cannot be read until we have learnt the language and become familiar with the characters in which it is written. It is written in mathematical language, and the letters are triangles, circles and other geometrical figures, without which means it is humanly impossible to comprehend a single word. ([46], 171)

Evariste Galois (1811–1832)

Since the beginning of the century, computational procedures have become so complicated that any progress by those means has become impossible, without the elegance which modern mathematicians have brought to bear on their research, and by means of which the spirit comprehends quickly and in one step a great many computations. It is clear that elegance, so vaunted and so aptly named, can have no other purpose... Go to the roots, of these calculations! Group the operations. Classify them according to their complexities rather than their appearances! This, I believe, is the mission of future mathematicians. This is the road on which I am embarking in this work. [47]

Lars Gårding (1919–)

I will argue below that creativity in mathematics does not differ from intellectual creativity in general. In the first place the creator must have what I call a net. This is a connected collection of facts, results, guesses and so on that the creator keeps in his head and to which he has immediate access. To build a net requires time, work, and interest in and love of the material. A creator must live with his net and more or less think of it all the time. The state of the net depends on the brain capacity and quickness of thought of the creator. Thinking about the net can be precise, dreamlike or haphazard or systematic or in pictures or not in pictures. All the ways of the human brain may be useful. A net is necessary for a creator but not sufficient. A net offers its owner the possibility to create something new and thus become a true creator. The above applies to engineers, philosophers, physicists, chemists, writers, artists and so on. Creativity in mathematics is considered to be mysterious by most people because they cannot imagine what a mathematical net could contain. Most of them even shy away from the opportunity to create a small net from the mathematics taught in the schools. What I have written here is true but unfortunately not scientific. [ν]

Carl Friedrich Gauss

Carl Friedrich Gauss (1777–1855)

Finally, two days ago, I succeeded, not on account of my painful efforts, but by the grace of God. Like a sudden flash of lightning, the riddle happened to be solved. I myself cannot say, what was the conducting thread, which connected what I previously knew, with what made my success possible. ([104], 89)

If others would but reflect on mathematical truths as deeply and as continuously as I have, they would make my discoveries. ([15], 326)

It is not knowledge, but the act of learning, not possession but the act of getting there, which grants the greatest enjoyment. When I have clarified and exhausted a subject, then I turn away from it, in order to go into darkness again; the never-satisfied man is so strange if he has completed a structure, then it is not in order to dwell in it peacefully, but in order to begin another. I imagine the world conqueror must feel thus, who, after one kingdom is scarcely conquered, stretches out his arms for others. ([36], 416)

The higher arithmetic presents us with an inexhaustible store of interesting truths—of truths, too, which are not isolated, but stand in a close internal connexion, and between which, as our knowledge increases, we are continually discovering new and sometimes wholly unexpected ties. A great part of its theories derives an additional charm from the peculiarity that important propositions, with the impress of simplicity upon them, are often easily discoverable by induction, and yet are of so profound a character that we cannot find their demonstration till after many vain attempts, and even then, when we do succeed, it is often by some tedious and artificial process, while the simpler methods may long remain concealed. [41]

Ennio De Giorgi (1928–1996)

I think that the origin of creativity in all fields is that which I call the capacity or disposition to dream: to imagine different worlds, different things, and to seek to combine them in one's imagination in various ways. To this ability—very similar in all the disciplines—one must add the ability to communicate those dreams unambiguously, requiring knowledge of the language and internal rules of the various disciplines. I believe this must be an ability to dream in an uncompartmentalized way, in the way called philosophy in antiquity. So, for the love of knowledge and confidence in communicating one's dreams unambiguously we must study the various languages, the differing theories of the various disciplines, and even of the arts—all the forms of human knowledge. ([42], 1099)

Invention and discovery have much in common: both come from searching. By discovery we mean "pulling the cover off" something which is already there, bringing to light something which was hidden; by invention we mean a construction out of that which seems to be lying about. The issue of whether a particular new idea was uncovered or constructed often cannot be resolved. At the bottom is the eternal issue: what does it mean to recognize something, to know it? What is invention, what is discovery? ... I think that proof is an invention—a construction of a road leading to the theorem. It happens sufficiently frequently that two mathematicians prove in independent ways the same theorem as stated, and the proof is rarely the same proof. Thus a theorem is something discovered; its proof is something invented. ([42], 1101)

James Whitbread Lee Glaisher (1848–1928)

In other branches of science, where quick publication seems to be so much desired, there may possibly be some excuse for giving to the world slovenly or ill-digested work, but there is no excuse in mathematics. The form ought to be as perfect as the substance, and the demonstrations as rigorous as those of Euclid. The mathematician has to deal with the most exact facts of Nature, and he should spare no effort to render his interpretation worthy of his subject, and to give to his work its highest degree of perfection. "*Pauca sed matura*" [few but ripe] was Gauss' motto. ([48], 467)

The mathematician requires tact and good taste at every step of his work, and he has to learn to trust to his own instinct to distinguish between what is really worthy of his efforts and what is not; he must take care not to be the slave of his symbols, but always to have before his mind the realities which they merely serve to express. ([48], 467)

Andrew Gleason (1921–2008)

It is notoriously difficult to convey a proper impression of the frontiers of mathematics to non-specialists... Ultimately the difficulty stems from the fact that mathematics is an easier subject than the other sciences... Consequently, many of the important primary problems of the subject—that is, problems which can be understood by an intelligent outsider—have either been solved or carried to a point where an indirect approach is clearly required. The great bulk of pure mathematical research is concerned with secondary, tertiary or higher-order problems, the very statement of which can hardly be understood until one has mastered a great deal of technical mathematics. ([5], 95)

Andrew Gleason

I've very much given to being gripped by explicit things. Sometimes little things, sometimes big things. Most of my work has been in response to very explicit, easily stated things. I'm very fond of problems in which somehow an at least very simple sounding hypothesis is sufficient to really pinch something together and make something out of it. ([5], 93)

Often enough mathematicians have been caught off base with some pathological wrinkle in some funny function or something so, you know, you do have to do the proof, but still in the end that much more important function of a proof in my view is to figure out why it works. ([5], 97)

Kurt Gödel (1906–1978)

But, despite their remoteness from sense experience, we do have something like a perception of the objects of set theory, as is seen from the fact that the axioms force themselves upon us as being true. I don't see any reason why we should have less confidence in this kind of perception, i.e., in mathematical intuition, than in sense perception. ([49], 484)

Ronald Graham (1935–)

I look at mathematics pretty globally. It represents the ultimate structure and order. And I associate doing mathematics with control. Jugglers like to be able to control a situation. There's a well-known saying in juggling: 'The trouble is that the balls go where you throw them.' It's just you. It's not the phases of the moon or someone else's fault. It's like chess. It's all out in the open. Mathematics is really there, for you to discover. ([64], 56)

Ulf Grenander (1923–)

...changing perspective, look at the problem from different angles. Search for similarities with other scientific work. Progress has often occurred at the boundary of two or more disciplines. [ν]

Chance has played a minor role. The main reason for occasional success is perseverance; never give up on a problem, continue day after day, week after week...also when it looks hopeless. [ν]

My attitude to mathematics has changed radically since I was a student. At that time I thought of mathematics as a body of theorems, a static concept. I learned later to look at it as a problem solving activity. The theorems are still important, but perhaps less so nowadays. [ν]

I have also learnt that it is important for young mathematicians to work on his/her own problems as early as possible ('mathematics is a young man's game'). Knowledge is needed but not enough, so that book learning should not be emphasized too much. [ν]

Alexandre Grothendieck (1928–)

In the work of discovery, this intense attention, this ardent solicitude, are an essential force, just like the warmth of the sun for the obscure gestation of seeds covered in nourishing soil, and for their humble and miraculous blossoming in the light of day. ([50], 49)

[I]f there is one thing in mathematics that fascinates me more than anything else (and doubtless always has), it is neither "number" nor "size", but always *form*. And among the thousand-and-one faces whereby form chooses to reveal itself to us, the one that fascinates me more than any other and continues to fascinate me, is *the structure* hidden in mathematical things. ([50], 27)

Alexandre Grothendieck

Richard Guy (1916–)

The human brain is a remarkable thing and we are a long way from understanding how it works. For most mathematical problems, immediate thought and pencil and paper—the usual things one associates with solving mathematical problems—are just totally inadequate. You need to

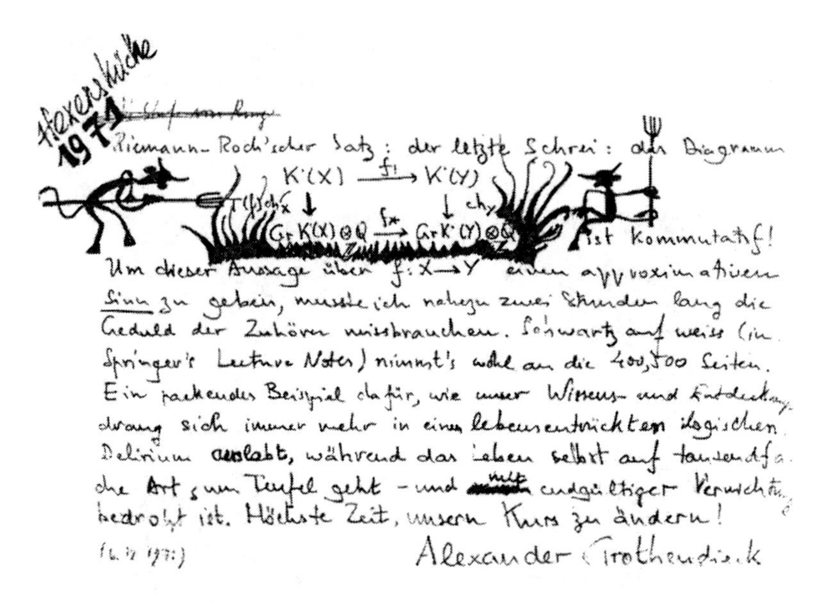

Figure 5. At the Universität Bielefeld, Grothendieck wrote this abstract into the colloquium book after he spoke there in 1971: "Witch's Kitchen 1971. Riemann-Roch Theorem: The 'dernier cri': The diagram [displayed] is commutative! To give an approximate sense to the statement about f: X ⇒ Y, I had to abuse the listeners' patience for almost two hours. A gripping example of how our thirst for knowledge and discovery indulges itself more and more in a logical delirium far removed from life, while life itself is going to Hell in a thousand ways—and is under the threat of final extermination. High time to change our course!"

understand the problem, make a few symbols on paper and look at them, and draw a few sausages on the paper. Most of us, as opposed to Erdős who would probably give an answer to a problem almost immediately, would then probably have to go to bed and, if we're lucky, when we wake up in the morning, we would already have some insight into the problem. On those rare occasions when I have such insight, I quite often don't know that I have it, but when I come to work on the problem again, to put pencil to paper, somehow the ideas just seem to click together and the thing goes through. It is clear to me that my brain must have gone on, in an almost combinatorial way, checking the cases or doing an enormous number of fairly trivial arithmetical computations. It seems to know the way to go. ([6], 147)

Jacques Hadamard (1865–1963)

... there is hardly any completely logical discovery. Some intervention of intuition issuing from the unconscious is necessary at least to initiate the logical work. ([51], 112)

... the ideas chosen by my unconscious are those which reach my consciousness, and I see that they are those which agree with my aesthetic sense. ([51], 39)

Paul Halmos (1916–2006)

Perhaps the closest analogy is between mathematics and painting. The origin of painting is physical reality, and so is the origin of mathematics—but the painter is not a camera and the mathematician is not an engineer. The painter of "Uncle Sam Wants You" got his reward from patriotism, from increased enlistments, from winning the war—which is probably different from the reward Rembrandt got from a finished work. How close to reality painting (and mathematics) should be is a delicate matter of judgment. Asking a painter to "tell a concrete story" is like asking a mathematician to "solve a real problem." Modern painting and modern mathematics are far out—too far in the judgment of some. Perhaps the ideal is to have a spice of reality always present, but not to crowd it the way descriptive geometry, say, does in mathematics, and medical illustration, say, does in painting. ([53], 388)

Mathematics—this may surprise you or shock you some—is never deductive in its creation. The mathematician at work makes vague guesses, visualizes broad generalizations, and jumps to unwarranted conclusions. He arranges and rearranges his ideas, and he becomes convinced of their truth long before he can write down a logical proof. The conviction is not likely to come early—it usually comes after many attempts, many failures, many discouragements, many false starts. It often happens that months of work result in the proof that the method of attack they were based on cannot possibly work, and the process of guessing, visualizing, and conclusion-jumping begins again... The deductive stage, writing the result down, and writing down its rigorous proof are relatively trivial once the real insight arrives; it is more like the draftsman's work, not the architect's. ([53], 380)

Paul Halmos

I spent most of a lifetime trying to be a mathematician—and what did I learn? What does it take to be one? I think I know the answer: you have to be born right, you must continually strive to become perfect, you must love mathematics more than anything else, you must work at it hard and without stop, and you must never give up... To be a scholar of mathematics you must be born with talent, insight, concentration, taste, luck, drive and the ability to visualize and guess. ([52], 400)

Talk to a painter… and talk to a mathematician, and you'll be amazed at how similarly they react. Almost every aspect of the life and of the art of a mathematician has its counterpart in painting, and vice versa. Every time a mathematician hears "I could never make my checkbook balance" a painter hears "I could never draw a straight line"—and the comments are equally relevant and equally interesting. The invention of perspective gave the painter a useful technique, as did the invention of 0 to the mathematician. Old art is as good as new; old mathematics is as good as new. Tastes change, to be sure, in both subjects, but a twentieth century painter has sympathy for cave paintings and a twentieth century mathematician for the fraction juggling of the Babylonians. A painting must be painted and then looked at; a theorem must be printed and then read. The painter who thinks good pictures, and the mathematician who dreams beautiful theorems are dilettantes; an unseen work of art is incomplete. In painting and in mathematics there are some objective standards of good—the painter speaks of structure, line, shape, and texture, where the mathematician speaks of truth, validity, novelty, generality—but they are relatively the easiest to satisfy. Both painters and mathematicians debate among themselves whether these objective standards should even be told to the young—the beginner may misunderstand and overemphasize them and at the same time lose sight of the more important subjective standards of goodness. Painting and mathematics have a history, a tradition, a growth. Students, in both subjects, tend to flock to the newest but, except the very best, miss the point; they lack the vitality of what they imitate, because, among other reasons, they lack the experience based on the traditions of the subject. ([53], 389)

…the source of all great mathematics is the special case, the concrete example. It is frequent in mathematics that every instance of a concept of seemingly great generality is in essence the same as a small and concrete special case. ([52], 324)

The joy of suddenly learning a former secret and the joy of suddenly discovering a hitherto unknown truth are the same to me—both have the flash of enlightenment, the almost incredibly enhanced vision, and the ecstasy and euphoria of released tension. ([52], 3)

Mathematics is not a deductive science—that's a cliché. When you try to prove a theorem, you don't just list the hypotheses, and then start to reason. What you do is trial and error, experimentation, guesswork. ([52], 321)

William Hamilton (1805–1865)

The mathematical process in the symbolical method [i.e., the algebraical] is like running a railroad through a tunneled mountain, that in the ostensive [i.e., the geometrical] like crossing the mountain on foot. The former causes us, by a short and easy transit, to our destined point, but in miasma, darkness, and torpidity, whereas the latter allows us to reach it only after time and trouble, but feasting us at each turn with glances of the earth and of the heavens, while we inhale the pleasant breeze, and gather new strength at every effort we put forth. ([111], 22)

Hermann Hankel (1839–1873)

If we compare a mathematical problem with an immense rock, whose interior we wish to penetrate, then the work of the Greek mathematicians appears to us like that of a robust stonecutter, who, with indefatigable perseverance, attempts to demolish the rock gradually from the outside by means of hammer and chisel; but the modern mathematician resembles an expert miner, who first constructs a few passages through the rock and then explodes it with a single blast, bringing to light its inner treasures. ([54], 9)

Isolated, so-called "pretty theorems" have even less value in the eyes of a modern mathematician than the discovery of a new "pretty flower" has to the scientific botanist, though the layman finds in these the chief charm of the respective sciences. ([54], 15)

G. H. Hardy (1877–1947)

The Theory of Numbers has always been regarded as one of the most obviously useless branches of Pure Mathematics. The accusation is one against which there is no valid defence; and it is never more just than when directed against the parts of the theory which are more particularly concerned with primes. A science is said to be useful if its development tends to accentuate the existing inequalities in the distribution of wealth, or more directly promotes the destruction of human life. The theory of prime numbers satisfies no such criteria. Those who pursue it will, if they are wise, make no attempt to justify their interest in a subject so trivial and so remote, and will console themselves with the thought that the greatest mathematicians of all ages have found in it a mysterious attraction impossible to resist. ([57], 350)

I have myself always thought of a mathematician as in the first instance an *observer*, a man who gazes at a distant range of mountains and notes down his observations. His object is simply to distinguish clearly and notify to others as many different peaks as he can. ([56], 59)

...there is, strictly, no such thing as mathematical proof; that we can...do nothing but *point*; that proofs are what Littlewood and I call *gas*, rhetorical flourishes designed to affect psychology, pictures on the board in the lectures, devices to stimulate the imagination of pupils. ([56], 59)

G.H. Hardy

The fact is that there are few more 'popular' subjects than mathematics. Most people have some appreciation of mathematics, just as most people can enjoy a pleasant tune; and there are probably more people really interested in mathematics than in music. Appearances may suggest the contrary, but there are easy explanations. Music can be used to stimulate mass emotion, while mathematics cannot; and musical incapacity is recognized (no doubt rightly) as mildly discreditable, whereas most people are so frightened of the name of mathematics that they are ready, quite unaffectedly, to exaggerate their own mathematical stupidity. ([55], 86)

...it is not disputed that mathematics is full of proofs, of undeniable interest and importance, whose purpose is not in the least to secure conviction. Our interest in these proofs depends on their formal and æsthetic properties. Our object is *both* to exhibit the pattern and to obtain assent. We cannot exhibit the pattern completely, since it is far too elaborate; and we cannot be content with mere assent from a hearer blind to its beauty. ([56], 59)

It is a melancholy experience for a professional mathematician to find himself writing about mathematicians. The function of a mathematician is to do something, to prove new theorems, to add to mathematics, and not to talk about what he or other mathematicians have done. ([55], 61)

In both [Euclid's and Pythagoras's] theorems (and in the theorems, of course, I include the proofs) there is a very high degree of *unexpectedness*, combined with *inevitability* and *economy*. The arguments take so odd and surprising a form; the weapons used seem so childishly simple when compared with the far-reaching results; but there is no escape from the conclusions... We do not want many 'variations' in the proof of a mathematical theorem: 'enumeration of cases', indeed, is one of the duller forms of mathematical argument. A mathematical proof should resemble a simple and clear-cut constellation, not a scattered cluster in the Milky Way. ([55], 113)

A chess problem is genuine mathematics, but it is in some way 'trivial' mathematics. However ingenious and intricate, however original and surprising the moves, there is something essential lacking. Chess problems are *unimportant*. The best mathematics is *serious* as well as beautiful—'important' if you like, but the word is very ambiguous, and 'serious' expresses what I mean much better. ([55], 88–89)

I am interested in mathematics only as a creative art. ([55], 115)

The mathematician's patterns, like the painter's or the poet's, must be beautiful; the ideas, like the colours or the words, must fit together in a harmonious way. Beauty is the first test: there is no permanent place in the world for ugly mathematics. ([55], 85)

I believe that mathematical reality lies outside us, that our function is to discover or *observe* it, and that the theorems which we prove, and which we describe grandiloquently as our 'creations', are simply the notes of our observations. This view has been held, in one form or another by many philosophers of high reputation, from Plato onwards, ... ([55], 123–4)

There is no scorn more profound, or on the whole more justifiable, than that of the men who make for the men who explain. Exposition, criticism, appreciation, is work for second-rate minds. ([55], 61)

A mathematician, like a painter or poet, is a maker of patterns. If his patterns are more permanent than theirs, it is because they are made with ideas. ([55], 84)

Harish-Chandra (1923–1983)

In mathematics there is an empty canvas before you which can be filled without reference to external reality. ([91], 202)

I have often pondered over the roles of knowledge or experience, on the one hand, and imagination or intuition, on the other, in the process of discovery. I believe that there is a certain fundamental conflict between the two, and knowledge, by advocating caution, tends to inhibit the flight of imagination. Therefore, a certain naiveté, unburdened by conventional wisdom, can sometimes be a positive asset. ([91], 206)

Charles Hermite (1822–1901)

There exists, if I am not mistaken, an entire world which is the totality of mathematical truths, to which we have access only with our mind, just as a world of physical reality exists, the one like the other independent of ourselves, both of divine creation. ([29], 46)

Charles Hermite

We are servants rather than masters in Mathematics. ([51], xii)

David Hilbert (1862–1943)

The infinite has always stirred the *emotions* of mankind more deeply than any other question; the infinite has stimulated and fertilized reason as few other *ideas* have; but also the infinite, more than any other *notion*, is in need of *clarification*. ([61], 371)

...there is no unsolvable problem at all. In place of the foolish *Ignorabimus* is, in contrast, our slogan:

We must know
We shall know. ([59], 127)

Who of us would not be glad to lift the veil behind which the future lies hidden; to cast a glance at the next advances of our science and at the secrets of its development during future centuries? What particular goals will there be toward which the leading mathematical spirits of coming generations will strive? What new methods and new facts in the wide and rich field of mathematical thought will the new centuries disclose? ([60], 437)

David Hilbert

A mathematical problem should be difficult in order to entice us, yet not completely inaccessible, lest it mock at our efforts. It should be to us a guide post on the mazy paths to hidden truths, and ultimately a reminder of our pleasure in the successful solution. ([60], 438)

In mathematics, as in any scientific research, we find two tendencies present. On the one hand, the tendency towards *abstraction* seeks to crystallize the *logical* relations inherent in the maze of material that is being studied, and to correlate the material in a systematic and orderly manner. On the other hand, the tendency towards *intuitive understanding* fosters a more immediate grasp of the objects one studies, a live *rapport* with them, so to speak, which stresses the concrete meaning of their relations. ([58], iii)

The appearance of what we call intrinsic harmony is also striking, in a sense other than that used by Leibniz, that it is an embodiment and realization of mathematical thought... We can only understand this agreement between nature and thought, between experiment and theory, if we take into consideration the formal component of both sides of nature and our understanding, and the mechanism on which it depends. The mathematical process of analysis gives us, or so it appears, the focus and footings to which matter in the real world, as well as thought in the world of the mind, withdraw and cede control and direction. ([59], 121)

The tool which governs the mediation between theory and practice, between thought and observation is mathematics; it builds the bridge and carries more and more of the load. It thereby happens that the basis of our entire present day culture, in so far as it is based on investigations dealing with nature, can be found in mathematics. ([59], 125)

Mathematical science is in my opinion an indivisible whole, an organism whose vitality is conditioned upon the connection of its parts. For with all the variety of mathematical knowledge, we are still clearly conscious of the similarity of the logical devices, the *relationship* of the *ideas* in mathematics as a whole and the numerous analogies in its different departments. We also notice that, the farther a mathematical theory is developed, the more harmoniously and uniformly does its construction proceed, and unsuspected relations are disclosed between hitherto separated branches of the science. So it happens that, with the extension of mathematics, its organic character is not lost but manifests itself the more clearly. ([60], 478)

E. W. Hobson (1856–1933)

Who has studied the works of such men as Euler, Lagrange, Cauchy, Riemann, Sophus Lie, and Weierstrass, can doubt that a great mathematician is a great artist? The faculties possessed by such men, varying greatly in kind and degree with the individual, are analogous with those requisite for constructive art. Not every mathematician possesses in a specially high degree that critical faculty which finds its employment in the perfection of form, in conformity with the ideal of logical completeness; but every great mathematician possesses the rarer faculty of constructive imagination. ([62], 290)

Melvin Hochster (1943–)

Chance plays some role, not a major one. Insight is very important, while inspiration often occurs only after many, many weeks, months, or even years of hard thought. Therefore relentless tenacity is important. The imaginative use of analogies has played a strong role for me. [ν]

Peter J. Huber (1934–)

Learning the work of others of course means following their thoughts, as opposed to thinking oneself. However I usually try to see whether I can find my own proofs for the assertions before I dig into their details. [ν]

If you have an idea, develop it on your own for, say, two months, and only then check whether the results are known. The reasons are: (1) If you try to check earlier, you won't recognize your idea in the disguise under which it appears in the literature. (2) If you read the literature too carefully beforehand, you will be diverted into the train of thought of the other author and stop exactly where he ran into an obstacle. [ν]

Perhaps an oil explorations simile is more appropriate. First I would have a promising, brilliant idea (the aha event) which would induce me to drill. But the eureka event ("I found it!") at best would come hours or days later, if and when the oil would begin to gush forth. That the idea had been brilliant and not merely foolish would be clear only in retrospect, after attempts to verify and confirm it. And later on one tends to suppress and forget foolish ideas because they are embarrassing (but they are indispensable companions to the brilliant ones!). [ν]

The difference between Archimedes' age and ours is that in the meantime it has become difficult to have brilliant new ideas whose correctness becomes obvious before the bath water has become uncomfortably cold. [ν]

Things are more complicated. If I was stumped by a problem and seemed to walk around a solid smooth, blank wall, then I would consciously stuff the problem into my subconscious, do something entirely different, and hope for some revelation to surface in due time (it often did). [ν]

...I should say that my subconscious usually would present a novel way of attack; if it presented a ready-made "solution", it often was quite wrong. [ν]

Serendipity is very important, but it only works if the ground is suitably prepared. I guess the reason is that the "straightforward" discoveries are easy to find, even if they may need a little sweat, and thus have been found by people working in the field before you. [ν]

Philologists and historians usually accumulate notes on cards before beginning to write linearly. I do not think this works so well in mathematics. I worked in spirals. [ν]

I found that in order to do creative work, I had to be at it without interruption for at least a week at a time. [ν]

When I had a successful idea, I could not let loose and worked furiously. [ν]

When things had been settled and written up, I felt exhausted and empty, and itched until I had a new promising idea. [ν]

Charles Hutton (1737–1823)

The method of fluxions is probably one of the greatest, most subtle, and sublime discoveries of any age: it opens a new world to our view, and extends our knowledge, as it were, to infinity; carrying us beyond the bounds that seemed to have been prescribed to the human mind, at least infinitely beyond those to which the ancient geometry was confined. ([66], 525)

Victor Ivrii (1949–)

Urban legend: Einstein was asked if his ideas are coming during his meal, [he] answered that he does not have time for a meal when ideas come. The same here: at the moment of discovery nobody has time to think how he/she is thinking. [ν]

Carl Jacobi (1804–1851)

It is true that M. Fourier held the opinion that the principal aim of mathematics is public utility and the explanation of natural phenomena; but a philosopher like him should have known that the only end of science is the honor of the human mind, and that, in this respect, a question about numbers is worth as much as a question about the system of the world. ([81], 103)

Mathesis est scientia eorum quae per se clara sunt.
Mathematics is the science of what is clear by itself. ([85], 135)

Zu Archimedes kam ein wissbegieriger Jüngling,
Weihe mich, sprach er zu ihm, ein in die göttliche Kunst,
Die so herrliche Dienste der Sternenkunde geleistet,
Hinter dem Uranos noch einen Planeten entdeckt.
Göttlich nennst Du die Kunst, sie ist's, versetzte der Weise,
Aber sie war es, bevor noch sie den Kosmo erforscht,
Ehe sie herrliche Dienste der Sternenkunde geleistet,
Hinter dem Uranos noch einen Planeten entdeckt.
Was Du im Kosmos erblickst, ist nur der Göttlichen Abglanz,
In der Olympier Schaar thronet die ewige Zahl. ([86], 338)

To Archimedes came a youth eager for knowledge.
Teach me, O Master, he said, that art divine
Which has rendered so noble a service to the lore of the heavens,

And back of Uranus yet another planet revealed.
Truly, the sage replied, this art is divine as thou sayest,
But divine it was ere it ever the Cosmos explored,
Ere noble service it rendered the lore of the heavens
And back of Uranus yet another planet revealed.
What in the Cosmos thou seest is but the reflection of God,
The God that reigns in Olympus is Number Eternal. ([28], 179)

Philip E. B. Jourdain (1879–1919)

...the process of mathematical discovery is a living and a growing thing. ([70], 6)

Mark Kac (1914–1984)

Creativity in mathematics, a vast and fascinating subject in itself, is, fortunately, not a matter of merely taking "any hypothesis that seems amusing" and deducing its consequences. If it were, it could never generate the kind of fire, despair and triumph that shine through the beautiful letters between Bolyai the elder and his son. In urging his son to abandon the struggle with the Fifth Postulate, the father, himself a noted and respected mathematician, wrote: "I have traveled past all the reefs of the infernal Dead Sea and have always come back with a broken mast and a torn sail." To which, some time later, came a reply in triumph and elation: "out of nothing I have created a new and wonderful world!" ([71], 156)

Irving Kaplansky (1917–2006)

Certainly one thing is to look at the first case—the easiest case that you don't understand completely. That general theorem down the road—hopefully you'll get to it by and by. The second piece of advice: do examples. Do a million examples. I think there are shameful cases of people making silly and reckless conjectures just because they didn't take the trouble to look at the first few examples. A well-chosen example can teach you so much. ([5], 131)

Sometimes when you work through an example, you suddenly get an insight which you wouldn't have got if you'd just been working abstractly with the hypothesis of your future theorem. I guess both of these are obvious pieces of advice, but they are ignored more often than they should be... if the problem is worthwhile, give it a good try. Take months, maybe years if necessary, before you announce to the world, "This is as far as I can go. I'm quitting." It is disgraceful to give up before you have given it a good college try. ([5], 131)

The advice of Gauss: publish little but make it good. ([5], 131)

Irving Kaplansky (left)

Edward Kasner (1878–1955)

Mathematics is often erroneously referred to as the science of common sense. Actually, it may transcend common sense and go beyond either imagination or intuition. It has become a very strange and perhaps frightening subject from the ordinary point of view, but anyone who penetrates into it will find a veritable fairyland, a fairyland which is strange, but makes sense, if not common sense. ([73], 8)

There is a famous formula—perhaps the most compact and famous of all formulas—developed by Euler from a discovery of the French mathematician De Moivre: $e^{i\pi} + 1 = 0$ .... It appeals equally to the mystic, the scientist, the philosopher, the mathematician. ([73], 103)

In purging mathematical philosophy of metaphysics, there has been ... a real gain. No longer is mathematics to be looked upon as a key to the truth with a capital *T*. It may now be regarded as a woefully incomplete, though enormously useful, Baedeker in a mostly uncharted land. Some of the landmarks are fixed; some of the vast network of roads is made understandable; there are guideposts for the bewildered traveler. ([73], 360)

The great mathematicians have acted on the principle "*Devinez avant de démontrer*," and it is certainly true that almost all important discoveries are made in this fashion. ([72], 285)

Mathematics, unlike the music of the spheres, is man's own handiwork, subject only to the limitations imposed by the laws of thought. ([73], 359)

Cassius Jackson Keyser (1862–1947)

It is in the inner world of pure thought, where all *entia* dwell, where is every type of order and manner of correlation and variety of relationship, it is in this infinite ensemble of eternal verities whence, if there be one cosmos or many of them, each derives its character and mode of being,—it is there that the spirit of mathesis has its home and its life. Is it a restricted home, a narrow life, static and cold and grey with logic, without artistic interest, devoid of emotion and mood and sentiment? That world, it is true, is not a world of *solar* light, not clad in the colours that liven and glorify the things of sense, but it is an illuminated world, and over it all and everywhere throughout are hues and tints transcending *sense*, painted there by radiant pencils of *psychic* light, the light in which it lies. It is a silent world, and, nevertheless, in respect to the highest principle of art—the interpenetration of content and form, the perfect fusion of mode and meaning—it even surpasses music. In a sense, it is a static world, but so, too, are the worlds of the sculptor and the architect. The figures, however, which reason constructs and the mathematic vision beholds, transcend the temple and the statue, alike in simplicity and in intricacy, in delicacy and in grace, in symmetry and in poise. Not only are this home and this life thus rich in æsthetic interests, really controlled and sustained by motives of a sublimed and supersensuous art, but the religious aspiration, too, finds there, especially in the beautiful doctrine of invariants, the most perfect symbols of what it seeks—the changeless in the midst of change, abiding things in a world of flux, configurations that remain the same despite the swirl and stress of countless hosts of curious transformations. The domain of mathematics is the sole domain of certainty. There and there alone prevail the standards by which every hypothesis respecting the external universe and all observation and all experiment must be finally judged. It is the realm to which all speculation and all thought must repair for chastening and sanitation—the court of last resort, I say it reverently, for all intellection whatsoever, whether of demon or man or deity. It is there that mind as mind attains its highest estate, and the condition of knowledge there is the ultimate object, the tantalising goal of the aspiration, the *Anders-Streben*, of all other knowledge of every kind. ([77], 313–314)

When the late Sophus Lie ... was asked to name the characteristic endowment of the mathematician, his answer was the following quaternion: Phantasie, Energie, Selbstvertrauen, Selbstkritik. ([74], 31)

Mathematics is no more the art of reckoning and computation than architecture is the art of making bricks or hewing wood, no more than painting is the art of mixing colors on a palette, no more than the science of geology is the art of breaking rocks, or the science of anatomy the art of butchering. ([74], 29)

Such is the character of mathematics in its profounder depths and in its higher and remoter zones that it is well nigh impossible to convey to one who has not devoted years to its exploration a just impression of the scope and magnitude of the existing body of the science. An imagination formed by other disciplines and accustomed to the interests of another field may scarcely receive suddenly an apocalyptic vision of that infinite interior world. But how amazing and how edifying were such a revelation, if it only could be made. ([74], 6)

It is in the mathematical doctrine of Invariance, the realm wherein are sought and found configurations and types of being that, amid the swirl and stress of countless hosts of transformations remain immutable, and the spirit dwells in contemplation of the serene and eternal reign of the subtile law of Form, it is there that Theology may find, if she will, the clearest conceptions, the noblest symbols, the most inspiring intimations, the most illuminating illustrations, and the surest guarantees of the object of her teaching and her quest, an Eternal Being, unchanging in the midst of the universal flux. ([74], 42)

The great generalization [of hyperspace] has made it possible to enrich, quicken and beautify analysis with the terse, sensuous, artistic, stimulating language of geometry. On the other hand, the hyperspaces are in themselves immeasurably interesting and inexhaustibly rich fields of research. Not only does the geometrician find light in them for the illumination of otherwise dark and undiscovered properties of ordinary spaces of intuition, but he also discovers there wondrous structures quite unknown to ordinary space... It is by creation of hyperspaces that the rational spirit secures release from limitation. In them it lives ever joyously, sustained by an unfailing sense of infinite freedom. ([75], 83)

The belief that mathematics, because it is abstract, because it is static and cold and gray, is detached from life, is a mistaken belief. Mathematics, even in its purest and most abstract estate, is not detached from life. It is just the ideal handling of the problems of life, as sculpture may idealize a human figure or as poetry or painting may idealize a figure or a scene. Mathematics is precisely the ideal handling of the problems of life, and the central ideas of the science, the great concepts about which its stately doctrines have been built up, are precisely the chief ideas with which life must always deal and which, as it tumbles and rolls about them through time and space, give it its interests and problems, and its order and rationality. ([76], 645–646)

Ibn Khaldun (1332–1406)

Geometry enlightens the intellect and sets one's mind right. All of its proofs are very clear and orderly. It is hardly possible for errors to enter into geometrical reasoning, because it is well arranged and orderly. Thus, the mind that constantly applies itself to geometry is not likely to fall into error. In this convenient way, the person who knows geometry acquires intelligence. ([78], 130)

Omar Khayyam (1048–1122)

By the help of God and with His precious assistance I say that algebra is a scientific art. The objects with which it deals are absolute numbers and (geometrical) magnitudes which, though themselves unknown, are related to things which are known, whereby the determination of the unknown quantities is possible. Such a thing is either a quantity or a unique relation, which is only determined by careful examination...What one seaches for in the algebraic art are the relations which lead from the known to the unknown, to discover which is the object of algebra as stated above. ([79], 123)

Omar Khayyam depiction

Felix Klein (1849–1925)

If the activity of a science can be supplied by a machine, that science cannot amount to much, so it is said; and hence it deserves a subordinate place. The answer to such arguments, however, is that the mathematician, even when he is himself operating with numbers and formulas, is by no means an inferior counterpart of the errorless machine, "thoughtless thinker" of Thomae; but rather, he sets for himself his problems with definite, interesting, and valuable ends in view, and carries them to solution in appropriate and original manner. He turns over to the machine only certain operations which recur frequently in the same way, and it is precisely the mathematician—one must not forget this—who invented the machine for his own relief, and who, for his own intelligent ends, designates the tasks which it shall perform. ([82], 22)

Daniel Kleitman (1934–)

It often does happen that when you are stuck, doing other things and coming back to it later is often the wisest course. Relevant ideas do pop up in your mind when you are taking a shower, and can pop up as well even when you are sleeping (many of these ideas turn out not to work very well), or even when you are driving. [ν]

Thus while you can turn the problem over in your mind in all ways you can think of and try to use all the methods you can recall or discover to attack it, there is really no standard approach that will solve it for you. At some stage, if you are lucky, the right combination occurs to you, and you are able to check it and use it to put an argument together. [ν]

You must try and fail by deliberate efforts, and then rely on a sudden inspiration or intuition—or if you prefer—luck. [ν]

It is often worthwhile to talk informally with others about your work. When you try to explain your ideas something new can occur to you, that you can later exploit. Also it sometimes happens that one person will say something in itself quite useless, but that gets misinterpreted by someone else into something valuable or into an idea that can lead to something valuable. Thus any random thought from someone else has the potential of leading you to just the new idea you need to progress in your work. [ν]

At the level of words, there are really no new ideas. Good results do not come from inventing new words. And even at a somewhat higher level there are not really new ideas. It is extended combinations of ideas that can be new and can solve difficult problems. [ν]

In working on this problem and in general, mathematicians wander in a fog not knowing what approach or idea will work, or if indeed any idea will, until by good luck, perhaps some novel ideas, perhaps some old approaches, conquer the problem. [ν]

Mathematicians, in short, are typically somewhat lost and bewildered most of the time that they are working on a problem. Once they find solutions, they also have the task of checking that their ideas really work, and that of writing them up, but these are routine, unless (as often happens) they uncover minor errors and imperfections that produce more fog and require more work. [ν]

The human mind is not only a computer, but one that has programmed itself. The active parts...get devoted to particular uses in a way that is proportional somehow to the amount of thought that the individual puts into that kind of use. The more a person thinks about some area, the more neurons are devoted to it, and the pathways through them become wider and easier to traverse. Also it seems that tunnelling is possible from one "idea" to another. You look in your mind for something and come back with something else which may, with enough looking and some luck, be just what you need to fill the gap in your argument. [ν]

What mathematicians write thus bears little resemblance to what they do: they are like people lost in mazes who only describe their escape routes never their travails inside. Of course they come to enjoy being lost. If you are not lost you cannot experience the thrill of finding your way out. [ν]

Morris Kline

Morris Kline (1908–1992)

The tantalizing and compelling pursuit of mathematical problems offers mental absorption, peace of mind amid endless challenges, repose in activity, battle without conflict, 'refuge from the goading urgency of contingent happenings', and the sort of beauty changeless mountains present to senses tried by the present-day kaleidoscope of events. ([83], 470)

Konrad Knopp (1882–1957)

The essence of mathematics is in its freedom. Mathematical creations, whatever else they might be, are free children of God, i.e., the free creations of the mind, more or less without any tie to outward objects. And for this reason, i.e., because of its similarity to the inner creative experience, mathematics can be compared to the arts. ([84], 14)

Donald Knuth (1938–)

Neither the mathematician nor the computer scientist is bound by a study of nature. With a pencil and paper we can control exactly what we are working on. A theorem is true or it's false. The fact that we deal with man-made things is common both to mathematics and computer science, but the nature of the thought process is sufficiently different that there probably is a justification for considering them to be two different views of the world—two different ways of organizing knowledge abstractly that have some points in common, but in a way they have their own domains. ([4], 193)

Donald Knuth

Sofia Kovalevskaya (1850–1891)

I can work at the same time with literature and mathematics. Many persons that have not studied mathematics confuse it with arithmetic and consider it a dry and arid science. Actually this science requires great fantasy, and one of the first mathematicians of our century [Weierstrass] very correctly said that it is not possible to be a complete mathematician without having the soul of a poet. As far as I am concerned I could never decide whether I had a greater inclination towards mathematics or literature. Just as my mind would tire from purely abstract speculations, I would immediately be drawn to observations about life, about stories; at another time, contrarily, when life would begin to seem uninteresting and insignificant then the incontrovertible laws of science would draw me to them. It may well be that in either of these sphere I would have done much more had I devoted myself to it exclusively, but I nevertheless could never give up either one completely. ([177], 316)

Leopold Kronecker (1823–1891)

God made integers, all else is the work of man. ([162], 19)

Nos mathematici sumus isti veri poetae,
Sed quod fingimus nos et probare decet.

Poets in truth are we in mathematics,
But our creations also must be proved. ([86], 48)

Wolfgang Krull (1899–1971)

It really seems true that a special sixth sense is needed to understand mathematics. The few who possess this sense fling themselves passionately into the subject; the rest stay as far away from it as possible or consider it a necessary evil. ([87], 48)

Imre Lakatos (1922–1974)

On the face of it there should be no disagreement about mathematical proof. Everybody looks enviously at the alleged unanimity of mathematicians; but in fact there is a considerable amount of controversy in mathematics. Pure mathematicians disown the proofs of applied mathematicians, while logicians in turn disavow those of pure mathematicians. Logicists disdain the proofs of formalists and some intuitionists dismiss with contempt the proofs of logicists and formalists. ([88], 61)

Susan Landau (1955–)

There's a touch of the priesthood in the academic world, a sense that a scholar should not be distracted by the mundane tasks of day-to-day living. I used to have great stretches of time to work. Now I have research thoughts while making peanut butter and jelly sandwiches. Sure it's impossible to write down ideas while reading "Curious George" to a two-year-old. On the other hand, as my husband was leaving graduate school for his first job, his thesis advisor told him, "You may wonder how a professor gets any research done when one has to teach, advise students, serve on committees, referee papers, write letters of recommendation, interview prospective faculty. Well, I take long showers." ([89], 704)

Serge Lang (1927–2005)

Why [do I do mathematics]? Why do you compose a symphony or a ballad? I already told you why. Because it gives me chills in the spine.

That's why. But I did not say that you should also get them. That's freedom for you. ([90], 17)

In general, new results are discovered by intuition, proofs are discovered by intuition and finally they are written up according to a logical pattern. But don't confuse the two. It's the same as in literature: grammar and syntax are not literature. When you write a musical piece, you use notes, but the notes are not the music. To read a piece of music from the written text is not a substitute for hearing the piece in Carnegie Hall or elsewhere. Logic is the hygiene of mathematics, just as grammar and syntax are the hygiene of language—and even then! "Under the bam, under the boo, under the bamboo tree...", there isn't any grammar. It is the poetry, the musical effect of words, poetic allusions, aesthetic impressionism, and many other things. But whereas the beauty of poetry pales under translation, the beauty of mathematics is invariant under linguistic transformations. ([90], 18)

Pierre-Simon Laplace (1749–1827)

Leibniz saw in his binary arithmetic the image of Creation... He imagined that Unity represented God, and Zero the void; that the Supreme Being drew all beings from the void, just as unity and zero expressed all numbers in his system of numeration. This conception was so pleasing to Leibniz that he communicated it to the Jesuit, Grimaldi, president of the Chinese tribunal for mathematics, in the hope that this emblem of creation would convert the Emperor of China, who was very fond of the sciences. I mention this merely to show how the prejudices of childhood may cloud the vision even of the greatest men! ([28], 15)

The theory of probabilities is at bottom nothing but common sense reduced to calculus; it enables us to appreciate with exactness that which accurate minds feel with a sort of instinct for which ofttimes they are unable to account. ([92], 440)

Peter Lax (1926–)

I don't have a high degree of self awareness while I think. The best I can say is that the crucial step is to become interested in good problems, but I cannot define what good problems are, nor describe the process of becoming interested. [v]

... a German mathematician, Schottky, when he reached the age of seventy or eighty ... was asked: "To what do you attribute your creativity and productivity?" The question threw him into great confusion. Finally he said: "But gentlemen, if one thinks of mathematics for fifty years, one must think of something!" It was different with Hilbert. This is a story I heard from Courant... At his seventieth birthday he was asked what he attributed his great creativity and originality to. He had the answer immediately: "I attribute it to my very bad memory." He really had to reconstruct everything, and then it became something else, something better. So maybe that is all I should say. I am between these two extremes. Incidentally, I have a very good memory. ([134], 228)

Peter Lax

Marshall Stone wrote that the mathematical language is enormously concentrated, it is like haikus. And I thought I would take it one step further and actually express a mathematical idea by a haiku. ([135], 229)

Speed depends on size
Balanced by dispersion
Oh, solitary splendor.

I like to start with some phenomenon, the more striking the better, and then use mathematics to try to understand it... There's an aesthetic quality, yes, but if you try to pin that down, you are just begging the question. What is beautiful is purely subjective. Saying something is beautiful may be no different from saying that you have a feeling that something is important. You know, one of the complaints that pure mathematicians have against applied mathematicians is that it is ugly... Beauty is in the eye of the beholder. It's a poor guide, aesthetics is. You have to feel that what you are doing is beautiful but, after all, someone used to classical art regards modern art as horrible and ugly. ([5], 155)

Lucien Le Cam (1924–2000)

There are mathematicians who do mathematics simply because they like it—it's a work of art. Some people work very hard on problems that have no relation to anything else. ([5], 174)

Most people think of me as some sort of mathematician dealing in abstract and abstruse problems. This is funny since I have essentially no formal mathematical training. ([5], 178)

Gottfried Leibniz (1646–1716)

It is an extremely useful thing to have knowledge of the true origins of memorable discoveries, especially those that have been found not by accident but by dint of meditation. It is not so much that thereby history may attribute to each man his own discoveries and that others should be encouraged to earn like commendation, as that the art of making discoveries should be extended by considering noteworthy examples of it. ([94], 22)

The art of discovering the causes of phenomena, or true hypotheses, is like the art of deciphering, in which one ingenious conjecture often greatly shortens the road. [95]

André Lichnerowicz (1915–1998)

The scholar has devoted his life to research, but it is very rare that, in the advancing years, the spark continues to glow. In a scientific notebook of Pasteur, we find a marginal note: "On the whole, nothing for two years," and this simple note denotes the anguish, among scholars, of knowing whether the spark has definitely been extinguished or whether the gift of creating science is still available to him. ([96], 189)

Mathematics is not interested in the nature of things; it puts 'being' in parentheses; this gives it at the same time its power and ambivalence: it is radically non-ontological. ([96], 186)

A mathematician is first of all an artisan learning by throwing himself against his own spirit, a necessary humility. He dreams and is a bit of an artist...I believe that if my neurophysiologist colleagues took electro-encephalograms of mathematicians, they would discover no difference between those of a working mathematician and a composer of music... Mathematics carries a form of witness all that the spirit of humans have in common, since mathematics does not depend on a civilization or a culture. ([96], 186)

J. E. Littlewood (1885–1977)

The theory of numbers is particularly liable to the accusation that some of its problems are the wrong sort of questions to ask. I do not myself think the danger is serious; either a reasonable amount of concentration leads to new ideas or methods of obvious interest, or else one just leaves the problem alone. 'Perfect numbers' certainly never did any good, but then they never did any particular harm. ([97], 74)

I constantly meet people who are doubtful, generally without due reason, about their potential capacity [as mathematicians]. The first test is whether you got anything out of geometry. To have disliked or failed to get on with other [mathematical] subjects need mean nothing; much drill and drudgery is unavoidable before they can get started, and bad teaching can make them unintelligible even to a born mathematician. ([97], 23)

J. E. Littlewood

In presenting a mathematical argument the great thing is to give the educated reader the chance to catch on at once to the momentary point and take details for granted: his successive mouthfuls should be such as can be swallowed at sight; in case of accidents, or in case he wishes for once to check in detail, he should have only a clearly circumscribed little problem to solve (e.g. to check an identity: *two* trivialities omitted can add up to an *impasse*). The unpractised writer, even after the dawn of a conscience, gives him no such chance; before he can spot the point he has to tease his way through a maze of symbols of which not the tiniest suffix can be skipped. ([97], 49)

Paul Malliavin (1925–)

During the occupation of Paris by Hitler, Hadamard was fortunate enough to save his life by fleeing to the United States; he suffered also deeply from the First World War having all his sons die fighting in the French Army at Verdun in 1916. In 1943, in New-York, with the help of a US research grant; Hadamard was able to write a book... [ν]

When I learn other works I am immediately trying to reconstruct the main results by myself. As Jean Leray said, "Mathematics is not a dead letter which can be stored in libraries, it is a living thinking. [ν]

[Mathematicians] are more interested in doing mathematics than speaking about it. [ν]

Benoit Mandelbrot (1924–2010)

It happens that I knew Hadamard personally since an uncle of mine was his successor at Collège de France... In addition, please understand that the mathematicians you are "polling" were largely trained in the 1950s or 1960s. Among them I am very atypical and may be the person whose view of mathematics is farthest from the "norms" of yesterday and closest to Hadamard's. [ν]

My method of work shows no real change, in fact—against every cliché—a steady improvement that may still continue at age 77. [ν]

Benoit Mandelbrot

When there is a need to fully assimilate something, I must redo everything in my own way. [ν]

My principal discoveries have arisen "spontaneously." Also, nearly every one was perceived as a change of direction. [ν]

The topics were first reputed to wander all around. But every so often there was a spurt of after-the-fact "self-organization" and reorganization that affected future progress. [ν]

As to my creative process, the sole peculiar feature that is identifiable, significant and worthy reporting is the essential role the eye continues to play. Hadamard would have understood this very well. [ν]

Do I experience feelings of illumination? Rarely, except in connection with chance, whose offerings I treasure. In my wandering life between concrete fields and problems, chance is continually important in two ways. A chance reading or encounter has often brought an awareness of existing mathematical tools that were new to me and allowed me to return to old problems I was previously obliged to leave aside. In other cases, a chance encounter suggested that old tools could have new uses that helped them expand. [ν]

My way of reading mathematics has not changed much. I read very quickly, trying first to understand the key points, and only afterwards fill in details if needed. [ν]

My passion for the history of ideas is boundless and I go to endless lengths, not to hide the influences from which I benefited, but to understand and express them thoroughly. To me, the value of a thought combines its novelty and difficulty with the depth of its roots. The greatest thrill is to add to streams of ideas that already have a long and recognizable past. [ν]

James Maxwell (1831–1879)

Mathematicians may flatter themselves that they possess new ideas which mere human language is as yet unable to express. Let them make the effort to express these ideas in appropriate words without the aid of symbols, and if they succeed they will not only lay us laymen under a lasting obligation, but, we venture to say, they will find themselves very much enlightened during the process, and will even be doubtful whether the ideas as expressed in symbols had ever quite found their way out of the equations into their minds. ([100], 328)

> O WRETCHED race of men, to space confined!
> What honour can ye pay to him, whose mind
> To that which lies beyond hath penetrated?
> The symbols he hath formed shall sound his praise,
> And lead him on through unimagined ways
> To conquests new, in worlds not yet created.
>
> First, ye Determinants! in ordered row
> And massive column ranged, before him go,

To form a phalanx for his safe protection.
Ye powers to the nth roots of -1!
Around his head in ceaseless cycles run,
As unembodied spirits of direction.

And you, ye undevelopable scrolls!
Above the host wave your emblazoned rolls,
Ruled for the record of his bright inventions.
Ye Cubic surfaces! by threes and nines
Draw round his camp your seven-amd-twenty lines—
The seal of Solomon in three dimensions.

March on, symbolic host! with step sublime,
Up to the flaming bounds of Space and Time!
There pause, until by Dickinson depicted,
In two dimensions, we the form may trace
Of him whose soul, too large for vulgar space,
In n dimensions flourished unrestricted. ([23], 414)

Jerrold Marsden (1942–2010)

One of the most intense experiences I had actually turned out to be nonsense. It occurred in a dream in which I really thought that I got insight into a really hard problem. When I got up, I rushed to my desk to think and after an hour realized that it was all gibberish. But it was quite intense. I find, in general, ideas that come to me "in the shower" are more reliable than those that come to me in my sleep (which does not happen very often). [ν]

I can be talking to a colleague or my wife or eating breakfast and suddenly, like a voice from the blue, I get told what to do. Hard to explain. [ν]

I was not going after it—it just happened. This is the opposite of the view of our current grant system, that imagines that one knows in advance what you want to do and then you go out and do it—like building a house. Research, really good research is not like that at all. [ν]

It is not just chance, but rather inspiration in the presence of lots of surrounding information. The surrounding information is really crucial. [ν]

I assimilate the work of others best through personal contact and being able to question them directly. [ν]

There were little creative things done when I was a student and there was lots of enthusiasm and dedication. Now, being more mature, the creative spirit interestingly has increased and I do less of the "busy work", leaving that to students and postdocs (which is good for them). [ν]

Barry Mazur (1937–)

[N]umber theory has an annoying habit: the field produces, without effort, innumerable problems which have a sweet, innocent air about them, tempting flowers; and yet ... the quests for the solutions of these problems have been known to lead to the creation (from nothing) of theories which spread their light on all of mathematics, have been known to goad mathematicians on to achieve major unifications of their science, have been known to entail painful exertion in other branches of mathematics to make those branches serviceable. Number theory swarms with bugs, waiting to bite the tempted flower-lovers who, once bitten, are inspired to excesses of effort! ([101], 593)

We heard, or made up, the gossip that Chevalley adamantly REFUSED to draw pictures when he did his algebraic geometry. The idealism, the wild-eyed rigor and asceticism of this gesture fired our imagination: we also decided 'WE DRAW NO PICTURES!' ([120], 80)

Dusa McDuff (1945–)

The answer came in a flash, unexpectedly, while I was showering the next morning. I saw a picture of the solution, right there, waiting to be described. [ν]

In my principal discoveries I have always been thinking hard trying to understand some particular problem. Often it is just a hard slog, I go round arguments time and again seeking for a hole in my reasoning, or for some way to formulate the problem/structures I see. Gradually some insight builds and I get to "know" how things function. The new idea might come during such a session, but it can also occur while I am doing something else, reading a novel, doing the cooking. [ν]

My imagination is quite visual (though not as much as some others I know of). Other mathematicians have a feel for algebraic structure; actual equations turn them on—not me... [ν]

I find it very hard to learn the results of others these days unless they are very close to my own research interests. I used to be able to absorb things rather passively, just reading and going over the chains of ideas. Now I need to work out examples, specific instances of the new ideas to feel that I have any real understanding. Anything one creates oneself is much more immediate and real and so harder to forget. [ν]

Gel'fand amazed me by talking of mathematics as though it were poetry. He once said about a long paper bristling with formulas that it contained the vague beginnings of an idea which he could only hint at and which he had never managed to bring out more clearly. I had always thought of mathematics as being much more straightforward: a formula is a formula, and an algebra is an algebra, but Gel'fand found hedgehogs lurking in the rows of his spectral sequences! [ν]

I try to build/find structure and cohesion in things that I am looking at. I think math is a language; one sees things with some internal eye and needs to find a language to express this. [ν]

Henry McKean (1930–)

Bye and bye I see it, often quite suddenly, and realize that it's all quite simple, as mathematics properly understood must always be. [ν]

I would not call it inspiration, but it is a rapid coming into focus of prior work, both overt and covert. The overt work is much the same as it always was. The covert work (in bed, on the subway, in dreams) is harder now. At the age 71, I no longer have the energy, the stamina to do it properly, so now this part is like a subliminal worry, not very productive, but who knows? Perhaps it still helps in its way. [ν]

My usual mode is to jump in and compute (I cannot really think without a pen in hand). Then having computed fast and probably wrong, I find that this particular calculation would not have done what I wanted anyhow, so I throw it out and start over. [ν]

Sometimes I simply repeat what are at bottom the same stupidities for weeks, and though this looks useless on the face of it, I get familiar with the question and learn a few tricks. Of course I know already what I want to come out, mostly by analogy with old things of my own or others, and I'm looking for the mathematical mechanism that makes it work. [ν]

Povzner says: "Gel'fand cannot solve difficult problems. He only solves simple problems meaning that when he's finished, it's obvious". [ν]

Curtis McMullen (1958–)

However I would like to mention that I find it almost impossible to have a creative thought while sitting at a desk. To the extent I "discover" things it is almost always while walking or pacing. Of course I can "work things out" at a desk, or on the computer, but to really "turn things over in my mind" I have to walk around. Similarly I think Hadamard mentions somewhere that he "thinks with his legs". [ν]

Yves Meyer (1939–)

My style is mostly original. I loved to attack problems which seemed out of reach. You then have to build everything from scratch. It is like exploring a new continent. A feeling of wilderness. [ν]

Somehow the intense relaxation of the journey let my brain function correctly. Indeed I sometimes feel that my own intellectual tension prevents the brain from functioning adequately. [ν]

To my opinion, mathematical intense creativity is the output of a kind of long crisis where one accumulates enough intellectual tension. This tension is hopefully liberated by a discovery that seems to happen within a few seconds of time. It is a kind of vision (internal vision). A kind of mystical experience which produces a remarkable happiness. [ν]

Cathleen Morawetz (1923–)

Ah, there's no excitement to beat the excitement of proving a theorem! Until you find out the next day that it's wrong... I'll tell you, though, there is something about being a mathematician that is extremely difficult. One of my children put it this way: It's that you're on stage all the time. You can't fake or shift the subject of conversation and so on. That's very demanding of people. ([5], 238)

I find that I may have emphasized the need to escape from the devils of mathematics to embark on the pleasures of the real world. But it works both ways, and sometimes the devils of the real world drive one into the pleasures of studying mathematics. ([5], 238)

Cathleen Morawetz

L. J. Mordell (1888–1972)

Mathematical study and research are very suggestive of mountaineering. Whymper made seven efforts before he climbed the Matterhorn in the

1860's and even then it cost the life of four of his party. Now, however, any tourist can be hauled up for a small cost, and perhaps does not appreciate the difficulty of the original ascent. So in mathematics, it may be found hard to realise the great initial difficulty of making a little step which now seems so natural and obvious, and it may not be surprising if such a step has been found and lost again. ([103], 4)

Marston Morse (1892–1977)

I made the same mistake that artists have made since the time of the Greeks, and placed mathematics alongside of the arts as their *handmaiden*... But mathematics is the *sister*, as well as the *servant* of the arts and is touched with the same madness and genius. ([104], 85)

Most convincing to me of the spiritual relations between mathematics and music, is my own very personal experience. Composing in an amateurish way, I get exactly the same elevation from a prelude that has come to me at the piano, as I do from a new idea that has come to me in mathematics. ([104], 88)

...discovery in mathematics is not a matter of logic. It is rather the result of mysterious powers which no one understands, and in which the unconscious recognition of beauty must play an important part. Out of an infinity of designs a mathematician chooses one pattern for beauty's sake, and pulls it down to earth, no one knows how. Afterwards the logic of words and of forms sets the pattern right. Only then can one tell someone else. The first pattern remains in the shadows of the mind. ([104], 88)

...what is it that a mathematician wants as an artist. I believe that he wishes merely to understand and to create. He wishes to understand, simply, if possible—but in any case to understand; and to create, beautifully, if possible—but in any case to create. ([104], 90)

Mathematicians of today are perhaps too exuberant in their desire to build new logical foundations for everything. Forever the foundation and never the cathedral. ([104], 91)

As Dürer knew full well, there is a center and final substance in mathematics whose perfect beauty is rational, but rational "in retrospect."

The discovery which comes before, those rare moments which elevate man, and the searchings of the heart which come after are not rational. They are gropings filled with wonder and sometimes sorrow. ([104], 92)

Marston Morse

John von Neumann (1903–1957)

One expects a mathematical theorem or a mathematical theory not only to describe and to classify in a simple and elegant way numerous and *a priori* disparate special cases. One also expects "elegance" in its "architectural," structural makeup. Ease in stating the problem, great difficulty in getting hold of it and in all attempts at approaching it, then again some very surprising twist by which the approach, or some part of the approach, becomes easy, etc. Also, if the deductions are lengthy or complicated, there should be some simple general principle involved, which "explains" the complications and detours, reduces the apparent arbitrariness to a few simple guiding motivations, etc. These criteria are clearly those of any creative art, and the existence of some underlying empirical, worldly motif in the background—often in a very remote background—overgrown by aestheticizing developments and followed into a multitude of labyrinthine variants, all this is much more akin to the atmosphere of art pure and simple than to that of the empirical sciences. ([106], 183)

But still a large part of mathematics which became useful developed with absolutely no desire to be useful, and in a situation where nobody could possibly know in what area it would become useful: and there were no general indications that it even would be so... This is true of all science. Successes were largely due to forgetting completely about what one ultimately wanted, or whether one wanted anything ultimately; in refusing to investigate things which profit, and in relying solely on guidance by criteria of intellectual elegance... And I think it extremely instructive to watch the role of science in everyday life, and to note how in this area the principle of *laissez faire* has led to strange and wonderful results. ([107], 652)

As a mathematical discipline travels far from its empirical sources, or still more, if it is second and third generation only indirectly inspired by ideas coming from "reality", it is beset with very grave dangers. It becomes more and more purely aestheticizing, more and more purely *l'art pour l'art* ...there is a great danger that the subject will develop along the line of least resistance...will separate into a multitude of insignificant branches... ([106], 183)

A discussion of the nature of any intellectual effort is difficult *per se*—at any rate, more difficult than the mere exercise of that particular intellectual effort. It is harder to understand the mechanism of an airplane, and the theories of the forces which lift and which propel it, than merely to ride in it, to be elevated and transported by it or even to steer it. It is exceptional that one should be able to acquire the understanding of a process without having previously acquired a deep familiarity with running it, with using it, before one has assimilated it in an instinctive and empirical way. ([106], 172)

... much of the best mathematical inspiration comes from experience and that it is hardly possible to believe in the existence of an absolute immutable concept of mathematical rigor, dissociated from all human experience. ([106], 179–180)

Max Newman (1897–1984)

That a mathematical theory is a lasting object to believe in few can doubt. Mathematical language is difficult but imperishable. I do not believe that any Greek scholar of today can understand the idiomatic undertones of Plato's dialogues, or the jokes of Aristophanes, as thoroughly as mathematicians can understand every shade of meaning in Archimedes' works. ([109], 167)

Issac Newton (1643–1727)

I do not know what I may appear to the world, but to myself I seem to have been only like a boy playing on the seashore, and diverting myself in now and then finding a smoother pebble or a prettier shell than ordinary, whilst the great ocean of truth lay all undiscovered before me. ([22], 407)

If I have seen farther, it is by standing on the shoulder of giants. ([22], 407)

Issac Newton

Louis Nirenberg (1925–)

I love inequalities. So if somebody shows me a new inequality, I say, "Oh, that's beautiful, let me think about it," and I may have some ideas connected with it. ([68], 447)

That's the thing I try to get across to people who don't know anything about mathematics, what fun it is! One of the wonders of mathematics is you go somewhere in the world and you meet other mathematicians, and it's like one big family. This large family is a wonderful joy. ([68], 449)

George Papanicolaou (1943–)

I often find that a whole area of physics or applications becomes transparent to me almost immediately as soon as I understand the mathematical structure that it has, the hard mathematical questions that it poses. [ν]

I look at papers only after I have had some overall idea of a problem and then I do not look at details. [ν]

Now I hardly read books in the fields I know, and often I do not like the books I read (for example in financial math or imaging and random media) because very few of them really contribute to the subject. [ν]

When an idea comes up that solves a hard problem that has been with you for a while you just know it is IT. [ν]

In the three or four cases where a clear advance was made the degree to which the idea worked out as hoped for is a measure of its importance and the satisfaction that it gives. Sometimes it is the simplicity of the idea or the ultimate simplicity of the results it gives. [ν]

Sometimes after thinking about a problem a complete solution comes out as if it had been worked out in detail before. I am not sure how this happens, perhaps because some methods and tools do become second nature to us after a while. [ν]

Seymour Papert (1928–)

Mathematical work does not proceed along the narrow logical path of truth to truth to truth, but bravely or gropingly follows deviations through the surrounding marshland of propositions which are neither simply and wholly true nor simply and wholly false. ([112], 195)

Blaise Pascal (1623–1662)

For judgement is what goes with instinct, just as knowledge goes with mind. Intuition falls to the lot of judgement, mathematics to that of the mind. ([114], 184)

Those who are accustomed to judge by feeling have no understanding of matters involving reasoning. For they want to go right to the bottom of things at a glance, and are not accustomed to look for principles. The others, on the contrary, who are accustomed to reason from principles, have no understanding of matters involving feeling, because they look for principles and are unable to see things at a glance. ([114], 230)

Blaise Pascal

Benjamin Peirce (1809–1880)

Mathematics is not the discoverer of laws, for it is not induction; neither is it the framer of theories, for it is not hypothesis; but it is the judge over both, and it is the arbiter to which each must refer its claims; and neither law can rule nor theory explain without the sanction of mathematics. ([116], 97)

Charles Peirce (1839–1914)

...mathematics is distinguished from all other sciences except only ethics, in standing in no need of ethics. Every other science, even logic—logic, especially—is in its early stages in danger of evaporating into airy nothingness, degenerating, as the Germans say, into an arachnoid film, spun from the stuff that dreams are made of. There is no such danger for pure mathematics; for that is precisely what mathematics ought to be. ([118], 1781)

It is terrible to see how a single unclear idea, a single formula without meaning, lurking in a young man's head, will sometimes act like an obstruction of inert matter in an artery, hindering the nutrition of the brain, and condemning its victim to pine away in the fullness of his intellectual vigor and in the midst of intellectual plenty. ([117], 37)

Among the minor, yet striking characteristics of mathematics, may be mentioned the fleshless and skeletal build of its propositions; the peculiar difficulty, complication, and stress of its reasonings; the perfect exactitude of its results; their broad universality; their practical infallibility. ([118], 1779)

Roger Penrose (1931–)

There are things in mathematics for which the term 'discovery' is indeed much more appropriate than 'invention'... These are the cases where much more comes out of the structure than is put into it in the first place. One may take the view that in such cases the mathematicians have stumbled upon 'works of god.' However, there are other cases where the mathematical structure does not have such a compelling uniqueness, such as when, in the midst of a proof of some result, the mathematician finds the need to introduce some contrived and far from unique construction in order to achieve some very specific end. In such cases no more is likely to come out of the construction than was put into it in the first place, and the word 'invention' seems more appropriate than 'discovery'. These are indeed just 'works of man'. On this view, the true mathematical discoveries would, in a general way, be regarded as greater achievements or aspirations than would the 'mere' inventions. ([115], 126)

Roger Penrose

Charles Peskin (1946–)

I'm convinced that I do my best work while asleep. The evidence for this is that I often wake up with the solution to a problem, or at least with a clear idea of how to proceed to solve it. [ν]

Plutarch (120–46 BC), of Archimedes

...being perpetually charmed by a domestic siren, that is, his geometry, [Archimedes] neglected his meat and drink, and took no care of his person; that he was often carried by force to the baths, and when there he would make mathematical figures in the ashes, and with his finger draws lines upon his body, when it was anointed; so much was he transported with intellectual delight, such an enthusiasm in science. ([122], 247)

Henri Poincaré (1854–1912)

Mathematics have a triple aim. They must furnish an instrument for the study of nature. But that is not all: they have a philosophic aim and, I dare maintain, an esthetic aim. They must aid the philosopher to fathom the notions of number, of space, of time. And above all, their adepts find therein delights analogous to those given by painting and music. They admire the delicate harmony of numbers and forms; they marvel when a new discovery opens to them an unexpected perspective; and has not the joy they thus feel the esthetic character, even though the senses take no part therein? Only a privileged few are called to enjoy it fully, it is true, but is not this the case for all the noblest arts? ([125], 75–76)

If I may be allowed to continue my comparison with the fine arts, the pure mathematician who should forget the existence of the exterior world would be like a painter who knew how to harmoniously combine colors and forms, but who lacked models. His creative power would soon be exhausted. ([125], 79–80)

Henri Poincaré

Mathematicians attach a great importance to the elegance of their methods and of their results, and this is not mere dilettantism. What is it that gives us the feeling of elegance in a solution or a demonstration? It is the harmony of the diverse parts, their symmetry, and their happy adjustment; in a word, all that introduces order, all that gives them unity, that enables us to obtain a clear comprehension of the whole as well as of the parts. ([124], 30–31)

A scientist worthy of his name, above all a mathematician, experiences in his work the same impression as an artist; his pleasure is as great and of the same nature. ([98], 44)

Then I turned my attention to the study of some arithmetical questions apparently without much success and without a suspicion of any connection with my preceding researches. Disgusted with my failure, I went to spend a few days at the seaside, and thought of something else. One morning, walking along the bluff, the idea came to me, with just the same characteristics of brevity, suddenness and immediate certainty, the arithmetic transformations of indefinite ternary quadratic forms were identical with those of non-Euclidean geometry. ([51], 13–14)

Henri Poincaré

The genesis of mathematical discovery is a problem which must inspire the psychologist with the keenest interest. For this is the process in which the human mind seems to borrow least from the exterior world, in which it acts, or appears to act, only by itself and on itself, so that by studying the process of geometric thought we may hope to arrive at what is most essential in the human mind. ([124], 46)

It seems to me, then, as I repeat an argument I have learnt, that I could have discovered it. This is often only an illusion; but even then, even if I am not clever enough to create for myself, I rediscover it myself as I repeat it. ([124], 50)

We can understand that this feeling, this intuition of mathematical order, which enables us to guess hidden harmonies and relations, cannot belong to every one. ([124], 50)

What, in fact, is mathematical discovery? It does not consist in making new combinations with mathematical entities that are already known. That can be done by anyone, and the combinations that could be so formed would be infinite in number, and the greater part of them would be absolutely devoid of interest. Discovery consists precisely in not constructing useless combinations, but in constructing those that are useful, which are an infinitely small minority. Discovery is discernment, selection. ([124], 50–51)

It may appear surprising that sensibility should be introduced in connexion with mathematical demonstrations, which, it would seem, can only interest the intellect. But not if we bear in mind the feeling of mathematical beauty, of the harmony of numbers and forms and of geometric elegance. It is a real æsthetic feeling that all true mathematicians recognize, and this is truly sensibility. ([124], 59)

For a fortnight I had been attempting to prove that there could not be any function analogous to what I have since called Fuchsian functions. I was at that time very ignorant. Every day I sat down at my table and spent an hour or two trying a great number of combinations, and I arrived at no result. One night I took some black coffee, contrary to my custom, and was unable to sleep. A host of ideas kept surging in in my head; I could almost feel them jostling one another, until two of them coalesced, so to speak, to form a stable combination. When morning came, I had established the existence of one class of Fuchsian functions... ([124], 52–53)

How does it happen that there are people who do not understand mathematics? If the science invokes only the rules of logic, those accepted by all well-formed minds, if its evidence is founded on principles that are common to all men, and that none but a madman would attempt to deny, how does it happen that there are so many people who are entirely impervious to it? ([124], 46–47)

The scientist does not study nature because it is useful, he studies it because he delights in it, and he delights in it because it is beautiful. If nature were not beautiful, it would not be worth knowing, and if nature were not worth knowing, life would not be worth living. ([125], 8)

George Pólya (1887–1985)

You have to guess a mathematical theorem before you prove it. ([128], vi)

A great discovery solves a great problem, but there is a grain of discovery in the solution of any problem. Your problem may be modest; but if it challenges your curiosity and brings into play your inventive faculties, and if you solve it by your own means, you may experience the tension and enjoy the triumph of discovery. ([126], v)

It is like going into an unfamiliar hotel room late at night without knowing even where to switch on the light. You stumble around in the dark room, perceive confused black masses, feel one or the other piece of furniture as you are groping for the switch. Then, having found it, you turn on the light and everything becomes clear. ([127], 54)

Why should a mathematician care for plausible reasoning? His science is the only one that can rely on demonstrative reasoning alone. The physicist needs inductive evidence, the lawyer has to rely on circumstantial evidence, the historian on documentary evidence, the economist on statistical evidence. These kinds of evidence may carry strong conviction, attain a high level of plausibility, and justly so, but can never attain the force of a strict demonstration... Perhaps it is silly to discuss plausible grounds in mathematical matters. Yet I do not think so. Mathematics has two faces. Presented in finished form, mathematics appears as a purely demonstrative science, but mathematics in the making is sort of an experimental science. A correctly written mathematical paper is supposed to contain strict demonstrations only, but the creative work of the mathematician resembles the creative work of the naturalist: observation, analogy, and conjectural generalizations, or mere guesses, if you prefer to say so, play an essential role in both. A mathematical theorem must be guessed before it is proved. The idea of a demonstration must be guessed before the details are carried through. ([129], 739)

George Pólya

The inexpensive but nice hotel in which I lived at that time as a young instructor was situated next to the woods in which the city maintained footpaths, benches, and tables for the inconvenience of the promenaders and picnickers. I had then the habit of doing my mathematical work in an agreeable and healthy way in strolling through the woods. I carried paper and pencil and occasionally a few books. Sometimes I sat down at a table and scribbled a few formulas. Then I continued my leisurely walk in thinking about my problem until another table invited me to sit down and scribble a little more or look up something in a book. ([130], 165)

Alfred Pringsheim (1850–1941)

The true mathematician is always a good deal of an artist, an architect, yes, of a poet. Beyond the real world, though perceptibly connected with it, mathematicians have intellectually created an ideal world, which they attempt to develop into the most perfect of all worlds, and which is being explored in every direction. None has the faintest conception of this world, except he who knows it. ([131], 381)

The domain, over which the language of analysis extends its sway, is, indeed, relatively limited, but within this domain it so infinitely excels ordinary language that its attempt to follow the former must be given up after a few steps. The mathematician, who knows how to think in this marvelously condensed language, is as different from the mechanical computer as heaven from earth. ([131], 367)

Just as the mathematician is able to form an acoustic image of a composition which he has never heard played by merely looking at its score, so the equation of a curve, which he has never seen, furnishes the mathematician with a complete picture of its course. Yea, even more: as the score frequently reveals to the musician niceties which would escape his ear because of the complication and rapid change of the auditory impressions, so the insight which the mathematician gains from the equation of a curve is much deeper than that which is brought about by a mere inspection of the curve. ([131], 364)

Alfréd Rényi (1921–1970)

If I feel unhappy, I do mathematics to become happy. If I am happy, I do mathematics to keep happy.

Ha rossz kedvem van, matematizálok, hogy jó kedvem legyen. Ha jó kedvem van, matematizálok, hogy megmaradjon a jó kedvem. ([157], 210)

Kenneth Ribet (1948–)

I'd say that most of my significant work comes from working on definite problems that are posed to me by other people or that arise as I read articles and listen to lectures. I don't believe that anything has come spontaneously. Of course, there are times when I'm stuck on something and have a spontaneous moment of illumination as I'm doing something else (e.g., biking to the university). The moment of illumination comes only after a foundation of preliminary hard work. [ν]

There's a certain amount of serendipity involved. For example, you might be reading one paper that will give you an idea in a subject that you thought was rather far removed from the paper that you are reading. You might come upon a new insight as you're preparing a lecture that is meant to explain what you have already done. [ν]

When I work on something, I try to cobble together an argument based on familiarity with special cases, analogy with other situations—basically anything that might be useful. When I finally have what I think is a valid argument, I start pulling it apart: Why does it work? Can it be simplified? Generalized? What's the main point? When I read the work of others, I go directly into the latter mode. [ν]

In the proof of Fermat's Last Theorem, there was a need to sort out the origin of the key ideas to answer queries from the press and the general public. In that case, we really bent over backwards to try to find out who had done what first. In general, I'd say that mathematicians like to give credit to their predecessors; this is more true in mathematics than in fast-changing fields like physics. When mathematicians use a technique or an idea in their papers, they generally try to give accurate references to original sources. [ν]

When I used to think about a mathematical proposition, I'd ask myself why it was supposed to be true and try to think about the impossibility of constructing counterexamples. Now, I can ask about nature: what seems to be true? If I know a pattern, I can try to prove that the pattern is right. Another thing I'd say is that a lot of so-called insight in mathematics is based on analogy. As you grow older and have more experience, you can use the experience to make new "insights." Finally, as I have grown older, I've become fonder of the big picture. What am I really doing? Is it likely that my efforts will lead to a result. There's a bit less of intense rooting around to try to establish things. [ν]

Julia Robinson (1919–1985)

[Mathematicians form] a nation of our own without distinctions of geographical origins, race, creed, sex, age, or even time (the mathematicians of the past and you of the future are our colleagues too)—all dedicated to the most beautiful of the arts and sciences. ([135], 1492)

I think that I have always had a basic liking for the natural numbers. To me they are the one real thing. We can conceive of a chemistry that is different from ours, or a biology, but we cannot conceive of a different mathematics of numbers. What is proved about numbers will be a fact in any universe. ([5], 264)

Julia Robinson

I would say that my stubbornness has been to a great extent responsible for whatever success I have had in mathematics. But then it is a common trait among mathematicians. ([5], 265)

I have been told that some people think that I was blind not to see the solution myself when I was so close to it. On the other hand, no one else saw it either. There are lots of things, just lying on the beach as it were, that we don't see until someone else picks one of them up. Then we all see that one. ([5], 278)

What I really am is a mathematician. Rather than being remembered as the first woman this or that, I would prefer to be remembered, as a mathematician should, simply for the theorems I have proved and the problems I have solved. ([5], 280)

George Riemann (1826–1866)

If only I had the theorems! Then I should find the proofs easily enough. ([65], 487)

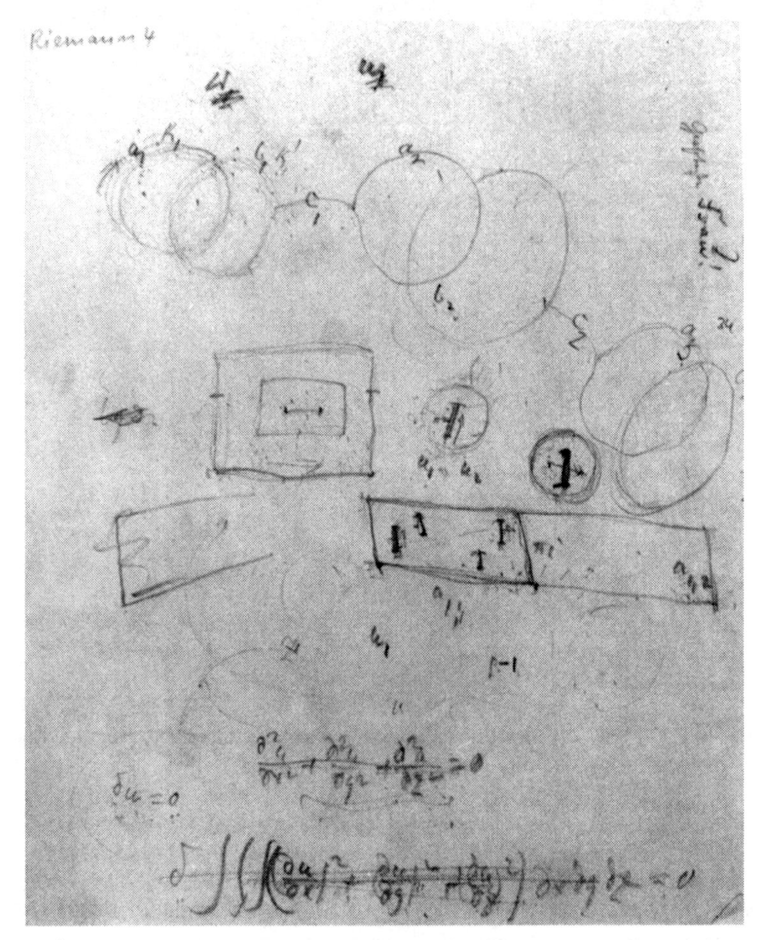

Figure 6. A page from the draft of Riemann's manuscript "Theorie der Abel'schen Functionen", 1857. ([102], 18)

Jakob Rosanes (1842–1922)

Everybody praises the incomparable power of the mathematical method, but so is everybody aware of its incomparable unpopularity. ([136], 60)

Gian-Carlo Rota (1932–1999)

What does a mathematician do when working on a mathematical problem? An adequate description of the project of solving a mathematical problem might require a thick volume. We will be content with recalling an old saying, probably going back to the mathematician George Pólya: "Few mathematical problems are ever solved directly." ([137], 99)

Every mathematician will agree that an important step in solving a mathematical problem, perhaps *the* most important step, consists of analyzing other attempts, either those attempts that have been previously carried out or attempts that he imagines might have been carried out, with a view to discovering how such "previous" approaches failed. In short, no mathematician will ever dream of attacking a substantial mathematical problem without first becoming acquainted with the *history* of the problem, be it the real history or an ideal history reconstructed by the gifted mathematician. ([137], 99)

Mary Ellen Rudin

Mary Ellen Rudin (1924–2013)

I say that there are lots of problems in mathematics that are interesting but have not been solved, and every time you solve one you think up a new one. Mathematics, therefore, is something that expands rather than contracts. And I tell that these questions are interesting just because you've followed a line of reasoning up to a certain point and the next natural thing to ask is what you're looking at. But that's really not explaining to them what kinds of things might be interesting to me. Sometimes that's pretty hard to explain—even to another mathematician. ([5], 301)

Mathematics is obviously something that women should be able to do very well. It's very intuitive. You don't need a lot of machinery, and you don't need a lot of physical strength. You just need stamina, and women often have a great deal of stamina. ([5], 301)

Bertrand Russell (1872–1970)

Mathematics, rightly viewed, possesses not only truth, but supreme beauty—a beauty cold and austere, like that of sculpture, without appeal to any part of our weaker nature, without the gorgeous trappings of painting or music, yet sublimely pure, and capable of a stern perfection such as only the greatest art can show. The true spirit of delight, the exaltation, the sense of being more than man, which is the touchstone of the highest excellence, is to be found in mathematics as surely as in poetry. ([138], 60)

I wanted certainty in the kind of way in which people want religious faith. I thought that certainty is more likely to be found in mathematics than elsewhere. But I discovered that many mathematical demonstrations, which my teachers expected me to accept, were full of fallacies, and that, if certainty were indeed discoverable in mathematics, it would be in a new field of mathematics, with more solid foundations than those that had hitherto been thought secure. But as the work proceeded, I was continually reminded of the fable about the elephant and the tortoise. Having constructed an elephant upon which the mathematical world could rest, I found the elephant tottering, and proceeded to construct a tortoise to keep the elephant from falling. But the tortoise was no more secure than the elephant, and after some twenty years of very arduous toil, I came to the conclusion that there was nothing more that I could do in the way of making mathematical knowledge indubitable. ([139], 54)

Pure mathematics consists entirely of such asseverations as that, if such and such a proposition is true of *anything*, then such and such another proposition is true of that thing. It is essential not to discuss whether the first proposition is really true, and not to mention what the anything is of which it is supposed to be true... If our hypothesis is about *anything* and not about some one or more particular things, then our deductions constitute mathematics. Thus mathematics may be defined as the subject in which we never know what we are talking about, nor whether what we are saying is true. ([140], 84)

The proof of self-evident propositions may seem, to the uninitiated, a somewhat frivolous occupation. To this we might reply that it is often by no means self-evident that one obvious proposition follows from another obvious proposition; so that we are really discovering new truths when we prove what is evident by a method which is not evident. But a more interesting retort is, that since people have tried to prove obvious propositions, they have found that many of them are false. Self-evidence is often a mere will-o'-the-wisp, which is sure to lead us astray if we take it as our guide. ([140], 86)

Hermann Schubert (1848–1911)

The intrinsic character of mathematical research and knowledge is based essentially on three properties: first, on its conservative attitude towards the old truths and discoveries of mathematics; secondly, on its progressive mode of development, due to the incessant acquisition of new knowledge on the basis of the old; and thirdly, on its self-sufficiency and its consequent absolute independence. ([142], 27)

...the three positive characteristics that distinguish mathematical *knowledge* from other knowledge ... may be briefly expressed as follows; first, mathematical knowledge bears more distinctly the imprint of truth on all its results than any other kind of knowledge; secondly, it is always a sure preliminary step to the attainment of other correct knowledge; thirdly, it has no need of other knowledge. ([142], 35)

Laurent Schwartz (1915–2002)

I would like to compare the phenomenon of sudden scientific discovery to the percolation of coffee. If water is poured over a mass of packed coffee grounds, at first it doesn't penetrate. Then some little trickles of water are born, but each one soon comes to a stop because the physical conditions don't allow it to continue. More and more little trickles begin, and they pierce farther and farther into the packed mass of grounds; they are not all born from the first trickle, but from collateral conditions arising from the existence of the first few trickles. And suddenly, one trickle makes a path right through the whole mass of packed grounds, and an artery is formed through which liquid passes quite quickly from the top to the bottom of the mass... Cerebral percolation is accompanied by an enormous quantity of subconscious work. Between the long period of reflection giving no result and the sudden discovery, the subconscious has done its work... It happens that the preliminary work preceding an important discovery does not give any publishable results, and no one notices them. If the person working on them doesn't keep them in mind, nothing will come of them. But the human spirit actually collects and hoards them. The author of the preliminary work may sometimes be the one who pierces through to the main discovery. Other times, it takes several mathematicians. ([143], 210–11)

Laurent Schwartz

Jean-Pierre Serre (1926–)

There are topics to which I come back from time to time … but I do not do this in a really systematic way. I rather follow my nose. As for flashes like the one Hadamard described, I have had only two or three in more than fifty years. They are wonderful … but much too rare! ([132], 210)

Jean-Pierre Serre

Lawrence Shepp (1936–2013)

One of my early mentors once asked me whether I ever got a great idea while having sex. I said that I had not observed this phenomenon, whereupon he said that he also never had observed this happening either. We had a good laugh over it. [ν]

How much did chance enter? It's a good question—it seems I got lucky quite often. Somehow I knew what I needed to know. It's a scary feeling. Sometimes it's hard to not believe in a greater presence, though the evidence is far from scientific. It has been said and I am paraphrasing, that "smart people are the ones who get lucky". [ν]

My work now at age 70 is a bit different; I still try to solve problems but I have moved to a different class of problems. I try to avoid getting bogged down with ones that need careful calculations which are more difficult for me to do now. I like to believe that my experience helps me to avoid problems that are too difficult, though it's hard to know whether I am simply "playing it safe". I have a long list of problems that I have been stuck on for many years and which may be too hard. But sometimes re-visiting them produces a solution. It is surely so that the most important thing for a scientist is to find "the right question". [ν]

Ideas come at strange times, once when I was unlocking the door to my home—I suddenly saw the entire solution to a problem I had been laboring over for a month or more. In another case I was having lunch with someone and jumped up off my chair more or less shouting, "I got it". In each of these cases, I was not aware that I was actually working on the problem but I suppose I must have been doing so somewhere in the back of my mind. Such "Eureka" moments produce a wonderful thrill, but is it innovation? One gets the same thrill when one solves a problem as a puzzle, even when one sees the idea in a chess or bridge problem or even a crossword answer. These are clearly not truly innovative since the solution is not at all original, except to oneself. [ν]

Yakov Sinai (1935–)

I have several papers on which I worked from two to three years and the solutions came as some kind of illuminating at very well defined moment of time. The concentration was so high, that it was difficult to write a detailed text afterwards. [ν]

Recently I needed one of my first papers, I was surprised when I recalled how much time I spent on it. Later it looked so simple. [ν]

I am very bad with reading others' papers. Either look trivial to me or I do not understand them at all. [ν]

Often I feel that I have a result but it is difficult for me to express it in a readable form. [ν]

Stephen Smale (1930–)

My work was mostly scribbling down ideas and trying to see how arguments could be put together. I would sketch crude diagrams of geometric objects flowing through space, and try to link the pictures with formal deductions. Deep in this kind of thinking and writing on a pad of paper, I was not bothered by the distractions of the beach. It was good to be able to take off from the research to swim. ([144], 861)

I'm not so loyal to mathematics as most mathematicians are... In many ways I'm different from most people. I'm not loyal to my subject. ([5], 310)

Beauty is very integrated with rarity... Beauty is connected so much with innovation and priority ... [In mathematics], it has to be something special to make it beautiful. If it's just ordinary, it's not beautiful. ([5], 320)

David Eugene Smith (1860–1944)

We study art because we receive pleasure from the great works of the masters, and probably we appreciate them the more because we have dabbled a little in pigments or in clay. We do not expect to be composers, or poets, or sculptors, but we wish to appreciate music and letters and the fine arts, and to derive pleasure from them and be uplifted by them. ([145], 16)

So it is with geometry. We study it because we derive pleasure from contact with a great and ancient body of learning that has occupied the attention of master minds during the thousands of years in which it has been perfected, and we are uplifted by it. To deny that our pupils derive this pleasure from the study is to confess ourselves poor teachers, for most pupils do have positive enjoyment in the pursuit of geometry, in spite of the tradition that leads them to proclaim a general dislike for all study. This enjoyment is partly that of the game,—the playing of a game that can always be won, but that cannot be won too easily. It is partly that of the æsthetic, the pleasure of symmetry of form, the delight of fitting things together. But probably it lies chiefly in the mental uplift that geometry brings, the contact with absolute truth, and the approach that one makes to the Infinite. We are not quite sure of any one thing in biology; our knowledge of geology is relatively very slight, and the economic laws of society are uncertain to every one except some individual who attempts to set them forth; but before the world was fashioned the square on the hypotenuse was equal to the sum of the squares on the other two sides of a right triangle, and it will be so after this world is dead; and the inhabitant of Mars, if he exists, probably knows its truth as we know it. The uplift of this contact with absolute truth, with truth eternal, gives pleasure to humanity to a greater or less degree, depending upon the mental equipment of the particular individual; but it probably gives an appreciable amount of pleasure to every student of geometry who has a teacher worthy of the name. ([145], 16–17)

David Eugene Smith

Henry J. Smith (1826–1883)

The bond of union among the physical sciences is the mathematical spirit and the mathematical method which pervades them.... Our knowledge of nature, as it advances, continuously resolves differences of quality into differences of quantity. All exact reasoning—indeed all reasoning—about quantity is mathematical reasoning; and thus as our knowledge increases, that portion of it which becomes mathematical increases at a still more rapid rate. ([146], 449)

Richard Stanley (1944–)

Chance is certainly a big factor… But insight/inspiration illumination is also important. Many of my main results required some sudden flash of insight. Most of my less important results did not involve such insight; it was more a case of determined plodding once the idea came to look at a certain problem. [ν]

In creating mathematics I spend much more time on pursuing wild ideas and performing crazy experiments than when I am learning the results of others. Probably 90% of my ideas amount to nothing, but one out of ten is more than sufficient to be a successful research mathematician. When learning the results of others, there is not the experience of being wrong most of the time or trying out ideas whose success is uncertain and in fact unlikely. [ν]

I am no longer able to concentrate as long and intensively as when I was younger. Part of the reason is that I am a lot busier in non-research related activities. I also have lots of undergraduates, graduate students, former students, and postdocs with whom I share ideas. Instead of working something out myself, I am just as likely to give it to a student. [ν]

Perhaps the key feature of the whole process is making connections. One has to recognize that two apparently disparate facts have something in common. Then comes the "99% perspiration" of nailing down exactly what is happening. [ν]

Shlomo Sternberg (1936–)

Mathematics is the science of order and mathematicians seek to identify instances of order and to formulate and understand concepts that enable us to perceive order in complicated situations. ([5], 94)

James Joseph Sylvester (1814–1897)

There are three ruling ideas, three so to say, spheres of thought, which pervade the whole body of mathematical science, to some one or other of which, or to two or all three of them combined, every mathematical truth admits of being referred; these are the three cardinal notions, of Number, Space and Order. ([153), 5)

...there is no study in the world which brings into more harmonious action all the faculties of the mind than [mathematics], ... or, like this, seems to raise them, by successive steps of initiation, to higher and higher states of conscious intellectual being... This accounts for the extraordinary longevity of all the greatest masters of the Analytic art, the Dii Majores of the mathematical Pantheon. Leibnitz lived to the age of 70; Euler to 76; Lagrange to 77; Laplace to 78; Gauss to 78; Plato, the supposed inventor of the conic sections, who made mathematics his study and delight, who called them the handles or aids to philosophy, the medicine of the soul, and is said never to have let a day go by without inventing some new theorems, lived to 82; Newton, the crown and glory of his race, to 85; Archimedes, the nearest akin, probably, to Newton in genius, was 75, and might have lived on to be 100, for aught we can guess to the contrary, when he was slain by the impatient and ill-mannered sergeant, sent to bring him before the Roman general, in the full vigour of his faculties, and in the very act of working out a problem; ... The mathematician lives long and lives young; the wings of his soul do not early drop off, nor do its pores become clogged with the earthy particles blown from the dusty highways of vulgar life. ([153], 657–658)
The world of ideas which [mathematics] discloses or illuminates, the contemplation of divine beauty and order which it induces, the harmonious connexion of its parts, the infinite hierarchy and absolute evidence of the truths with which it is concerned, these, and such like, are the surest grounds of the title of mathematics to human regard, and would remain unimpeached and unimpaired were the plan of the universe unrolled like a map at our feet, and the mind of man qualified to take in the whole scheme of creation at a glance. ([153], 659)

He who would know what geometry is, must venture boldly into its depths and learn to think and feel as a geometer. I believe that it is impossible to do this, and to study geometry as it admits of being studied and am conscious it can be taught, without finding the reasoning invigorated, the invention quickened, the sentiment of the orderly and beautiful awakened and enhanced, and reverence for truth, the foundation of all integrity of character, converted into a fixed principle of the mental and moral constitution, according to the old and expressive adage "*abeunt studia in mores*." ([150], 9)

I know, indeed, and can conceive of no pursuit so antagonistic to the cultivation of the oratorical faculty ... as the study of Mathematics. An eloquent mathematician must, from the nature of things, ever remain as rare a phenomenon as a talking fish, and it is certain that the more anyone gives himself up to the study of oratorical effect the less will he find himself in a fit state to mathematicize. It is the constant aim of the mathematician to reduce all his expressions to their lowest terms, to retrench every superfluous word and phrase, and to condense the Maximum of meaning into the Minimum of language. He has to turn his eye ever inwards, to see everything in its dryest light, to train and inure himself to a habit of internal and impersonal reflection and elaboration of abstract thought, which makes it most difficult for him to touch or enlarge upon any of those themes which appeal to the emotional nature of his fellowmen. When called upon to speak in public he feels as a man might do who has passed all his life in peering through a microscope, and is suddenly called upon to take charge of a astronomical observatory. He has to get out of himself, as it were, and change the habitual focus of his vision. ([151], 72–73)

May not Music be described as the Mathematic of sense, Mathematic as the Music of the reason? the soul of each the same! Thus the musician *feels* Mathematic, the mathematician *thinks* Music,—Music the dream, Mathematic the working life—each to receive its consummation from the other when the human intelligence, elevated to its perfect type, shall shine forth glorified in some future Mozart-Dirichlet or Beethoven-Gauss—a union already not indistinctly foreshadowed in the genius and labours of a Helmholtz! ([152], 419)

...we are told that "Mathematics is that study which knows nothing of observation, nothing of induction, nothing of causation." I think no statement could have been made more opposite to the undoubted facts of the case, that mathematical analysis is constantly invoking the aid of new principles, new ideas, and new methods, not capable of being defined by any form of words, but springing direct from the inherent powers and activity of the human mind, and from continually renewed introspection of that inner world of thought of which the phenomena are as varied and require as close attention to discern as those of the outer physical world (to which the inner one in each individual man may, I think, be conceived to stand in somewhat the same general relation of correspondence as a shadow to the object from which it is projected, or as the hollow palm of one hand to the closed fist which it grasps of the other), that it is unceasingly calling forth the faculties of observation and comparison, that one of its principal weapons is induction, that it has frequent recourse to experimental trial and verification, and that it affords a boundless scope for the exercise of the highest efforts of imagination and invention. ([153], 654)

P.G. Tait (1831–1901)

The flights of the imagination which occur to the pure mathematician are in general so much better described in his formulæ, than in words, that it is not remarkable to find the subject treated by outsiders as something essentially cold and uninteresting. ([155], 271)

Jean Taylor (1944–)

Insight to me is essential. I need to have an idea where things are likely to go. I think I learn the most, however, when my insight turns out to be lacking in some way, and I am surprised at the results. Again, it takes a lot of rigorous effort to turn an insight into something that could be called mathematics. [ν]

I think I just work very hard when I'm being productive. I seem to go into some kind of a trance; when someone talks to me when I'm in that state, I have to struggle to climb out and pay attention, and then it can take a while to get back into the groove. There are times when I just can't seem to have all the neurons firing that are needed for the problem, and there is no way to marshal them; I just have to wait until the next day. [ν]

People can have insights and inspirations, but if someone doesn't do the work of getting all the details down, it cannot become part of mathematics. Thus the interesting thing to me isn't necessarily the origins but the intense concentration that mathematics requires. [ν]

William Thomson (1824–1907)

Once when lecturing to a class he [Lord Kelvin] used the word "mathematician," and then interrupting himself asked his class: "Do you know what a mathematician is?" Stepping to the blackboard he wrote upon it:

$$\int_{-\infty}^{+\infty} e^{-x^2}\, dx = \sqrt{\pi}$$

Then putting his finger on what he had written, he turned to his class and said: "A mathematician is one to whom *that* is as obvious as that twice two makes four is to you. Liouville was a mathematician. ([156], 1139)

William Thurston (1946–2012)

I think of myself as learning the outskirts of mathematics. I think mathematics is a vast territory. The outskirts of mathematics are the outskirts of mathematical civilization. There are certain subjects that people learn about and gather together. Then there is a sort of inevitable development in those fields. You get to the point where a certain theorem is bound to be proved, independent of any particular individual, because it is just in the path of development. I enjoy trying to find mathematical topics that people haven't thought to think about. Then I work there. I like elbow room. ([5], 332)

...basic research in mathematics is detached from reality in its motivation. ([5], 333)

It's very time-consuming to write things down... I worry a lot. I try to avoid thinking about new subjects. But I still get behind. It's too easy to be distracted. There are too many interruptions. ([5], 333)

There is great power in truth and sincerity. The mathematics community has tremendous reserves of human potential energy. If we are lean and hungry, we are likely to use our energy. If we are honest, it is likely to be effective...([5], 334)

...mathematicians enjoy being in the ivory tower. I think that most mathematicians love mathematics for mathematics' sake. They really do like the feeling of being in an ivory tower. For the most part they are not motivated by applications. But I believe that, whatever their personal motivation is for doing mathematics, in most cases the mathematics they generate will ultimately have significant applications. The important thing is to do the mathematics. But, of course, it's important to have people thinking about applications too. ([5], 335)

William Thurston

Why do we try to prove things anyway? I think because we want to understand them. We also want a sense of certainty. Mathematics is a very deep field. Its results are stacked very high, and they depend on each other a lot. You build a tower of blocks but if one block is a bit wobbly, you can't build the tower very high before it will fall over. So I think mathematicians are concerned about rigor, which gives us certainty. That's one reason that we concentrate so much more on proof than do other scientists. But I also think proofs are so that we can understand. I guess I like explanations rather than step-by-step rigorous demonstrations. ([5], 340–41)

There are mathematicians, and then there's the rest of the world, and not much interaction between the two. ([5], 335)

The inner force that drives mathematicians isn't to look for applications; it's to understand the structure and inner beauty of mathematics. ([5], 335)

Personally I like to see lots of relations between lots of different things. I really enjoy that kind of integration you can have when you take very particular nitty-gritty questions and tie them together in very abstract theories. ([5], 336)

Very little [of mathematics] is easily accessible. But I think a lot more of it can be explained so that a lot more people understand it. On the level we're talking about. I like to try to make mathematics easy, not to make it hard. I think there is a tendency among mathematicians to try to make it hard. I try to combat that when I see people wrap up their mathematics in formal fancy theories that make it less accessible. ([5], 337)

I think that vision is somehow distracting to the spatial sense, because we have a spatial sense that is more than just vision. People associate it with vision, but it's not the same.... Sometimes pictures can get in the way. Sometimes one can evoke better pictures in one's head just by words. The spatial image is important, but it's what's in the head that counts. ([5], 337)

Alan Turing (1912–1954)

Mathematical reasoning may be regarded rather schematically as the exercise of a combination of two facilities, which we may call *intuition* and *ingenuity*. The activity of the intuition consists in making spontaneous judgements which are not the result of conscious trains of reasonings... The exercise of ingenuity in mathematics consists in aiding the intuition through suitable arrangements of propositions, and perhaps geometrical figures or drawings. It is intended that when these are really well arranged the validity of the intuitive steps which are required cannot seriously be doubted. The parts played by these two faculties differ of course from occasion to occasion, and from mathematician to mathematician. ([158], 135)

Messages from the Unseen World

III The Universe is the interior of the Light Cone of the Creation

IV Science is a Differential Equation. Religion is a Boundary Condition

Arthur Stanley

V Hyperboloids of wondrous Light
Rolling for aye through Space and Time
Harbour ~~Shelter~~ those Waves which somehow Might
Play out God's holy pantomime

VI Particles are founts

VII Charge = $\frac{e}{\pi}$ arg of character of a 2π rotation

VIII The Exclusion Principle is laid down purely for the benefit of the electrons themselves, who might be corrupted (and become dragons or demons) if allowed to associate too freely.

Figure 7. Three postcards from Alan Turing headed 'Messages from the Unseen World', an allusion to Eddington's 1929 book Science and the Unseen World. ([63], 512)

Karen Uhlenbeck (1942–)

You need to spend lots of time thinking about the problem. When I was young I had a great difficulty in sleeping, so I spent lots of time thinking about mathematical problems instead of sleeping. It is very difficult to build up intuition. Somehow you need to spend time with the problems and I am not sure how to teach that. ([159], 10)

Stanisław Ulam (1909–1984)

In many cases, mathematics is an escape from reality. The mathematician finds his own monastic niche and happiness in pursuits that are disconnected from external affairs. Some practice it as if using a drug. Chess sometimes plays a similar role. In their unhappiness over the events of this world, some immerse themselves in a kind of self-sufficiency in mathematics. (Some have engaged in it for this reason alone.) ([160], 120)

I recall a session with Mazur and Banach at the Scottish Café which lasted seventeen hours without interruption except for meals... There would be brief spurts of conversation, a few lines would be written on the table, occasional laughter would come from some of the participants, followed by long periods of silence during which we just drank coffee and stared vacantly at each other. The café clients at neighboring tables must have been puzzled by these strange doings. Is it such persistence and habit of concentration which somehow becomes the most important prerequisite for doing genuinely creative mathematical work. ([160], 33–34)

Stanisław Ulam

John Venn (1834–1923)

Without consummate mathematical skill, on the part of some investigators at any rate, all the higher physical problems would be sealed to us; and without competent skill on the part of the ordinary student no idea can be formed of the nature and cogency of the evidence on which the solutions rest. Mathematics are not merely a gate through which we may approach if we please, but they are the only mode of approach to large and important districts of thought. ([161], xix)

John Venn

Grace Wahba (1934–)

When I was younger I spent more time learning mathematics, but more recently I've spent more time interacting with students and colleagues. This might be a function of getting older, or, a function of the fact that the field of Statistics has changed a lot or a function of the fact that I think best standing at a whiteboard with a whiteboard marking pen in my hand: The part like pure mathematics (theorem: proof) plays a smaller role (or at least, smaller in my life), and numerical methods and innovative applications are playing a bigger role... [ν]

My only comment on the creative process (whatever that is), is that for me it has involved thinking about a problem, talking it over with others, in some more recent cases working with others who did a lot of computational work which provided insight, bumping into the right people at the right time, or, having a brilliant student, there's some luck involved (purely random encounters with Donald R. Johnson (at a committee meeting) and Steve Ackerman (at the Dane County Airport), for example) not to mention a certain amount of `chutzpah' believing that you can actually solve the problem with enough stick-to-it-iveness—with Sunduz, I heard her talk about a different problem and thought she could fill in some holes in our work and she was great. Another thing... in today's academic world with teaching loads, committee jobs, paperwork, paperwork, meetings, meetings—its pretty necessary to have at least some grant support to provide time to think. [ν]

Karl Weierstrass (1815–1897)

It is true that a mathematician, who is not somewhat of a poet, will never be a perfect mathematician. ([37], 149)

André Weil (1906–1998)

If logic is the hygiene of the mathematician, it is not his source of food; the great problems furnish the daily bread on which he thrives. ([166], 297)

Rigor is to the mathematician what morality is to man. It does not consist in proving everything, but in maintaining a sharp distinction between what is assumed and what is proved, and in endeavoring to assume as little as possible at every stage. ([164], 35)

André Weil

But if, like Panurge, we ask the oracle questions which are too indiscreet, then the oracle will answer as to Panurge: "Drink!" Advice which the mathematician is only too glad to follow, satisfied that he is to quench the thirst at the very sources of knowledge, satisfied that these sources always gush pure and abundant, while others must have recourse to the muddy paths of a sordid actuality. That if one reproaches him for his arrogant attitude, if one challenges him to engage himself in the actual world, if one asks why he persists on these high glaciers where none but others like him can follow him, he answers with Jacobi: "For the honor of the human spirit!" ([166], 306)

> Behold a collection of vectors.
> A field acts alone, abstract, commutative.
> The dual remains afar, solitary and plaintive,
> Searching for the isomorphism but finding it elusive.
>
> Suddenly, bilinear, a spark flew,
> Begetting the twice-distributive operator.
> In the product's net, all the captive vectors
> Will endlessly celebrate the more beautiful structure.
>
> But the basis disturbed this aeolian hymn:
> The exultant vectors have coordinates.
> Cartan knows not what to do and understands not a thing.
>
> And it's the end. Operators, vectors, ruined.
> A wretched matrix expires. The naked field
> Turns in on itself, embraced in its self-imposed laws.
>
> ([99], 113)

Every mathematician worthy of the name has experienced, if only rarely, the state of lucid exaltation in which one thought succeeds another as if miraculously, and in which the unconscious (however one interprets this word) seems to play a role. In a famous passage, Poincaré describes how he discovered Fuchsian functions in such a moment. About such states, Gauss is said to have remarked as follows: "*Procreare jucundum* (to conceive is a pleasure)"; he added, however, "*sed parturire molestum* (but to give birth is painful)." Unlike sexual pleasure, this feeling may last for hours at a time, even for days. Once you have experienced it, you are eager to repeat it but unable to do so at will, unless perhaps by dogged work which it seems to reward with its appearance. ([165], 91)

... strategy means the art of recognizing the main problems, attacking them at their weak points, setting up future lines of advance. Mathematical strategy is concerned with long-range objectives; it requires a deep understanding of broad trends and of the evolution of ideas over long periods. ([163], 205)

...the ability to recognize mathematical ideas in obscure and inchoate form, and to trace them under the many disguises which they are apt to assume before coming out in full daylight, is most likely to be coupled with a better than average mathematical talent. More than that, it is an essential component of such talent, since in large part the art of discovery consists in getting a firm grasp on the vague ideas which are "in the air," some of them flying all around us, some (to quote Plato) floating around in our own minds. ([163], 207)

Alan Weinstein (1943–)

Most of what I "discover" in my dreams turns out to be nonsense. [ν]

Hermann Weyl (1885–1955)

We must learn a new modesty. We have stormed the heavens, but succeeded only in building fog upon fog, a mist which will not support anybody who earnestly desires to stand upon it. What is valid seems so insignificant that it may be seriously doubted whether analysis is at all possible. ([28], 227)

By the mental process of thinking we try to ascertain truth; it is our mind's effort to bring about its own enlightenment by evidence. ([168], 68)

At the basis of all knowledge there lies: (1) *Intuition*, mind's originary act of "seeing" what is given to him... (2) *Understanding and expression*. Even in Hilbert's formalized mathematics I must understand the directions given me by communication in words for how to handle the symbols and formulas. (3) *Thinking the possible*. In science a very strin-

Hermann Weyl

gent form of it is exercised when, by thinking out the possibilities of the mathematical game, we try to make sure that the game will never lead to a contradiction; a much freer form is the imagination by which theories are conceived. ([168], 76)

The stringent precision attainable for mathematical thought has led many authors to a mode of writing which must give the reader an impression of being shut up in a brightly illuminated cell where every detail sticks out with the same dazzling clarity, but without relief. I prefer the open landscape under a clear sky with its depth of perspective, where the wealth of sharply defined nearby details gradually fades away towards the horizon. ([167], viii)

While in other fields brief allusions are met by ready understanding, this is unfortunately seldom the case with mathematical ideas. ([25], 84)

My work has always tried to unite the truth with the beautiful; but when I had to choose one or the other, I usually chose the beautiful. ([39], 458)

It is a fact that beautiful general concepts do not drop out of the sky. The truth is that, to begin with, there are definite concrete problems, with all their undivided complexity, and these must be conquered by individuals relying on brute force. Only then come the axiomatizers and conclude that instead of straining to break in the door and bloodying one's hands one should have first constructed a magic key of such and such shape and then the door would have opened quietly, as if by itself. But they can construct the key only because the successful breakthrough enables them to study the lock front and back, from the outside and from the inside. Before we can generalize, formalize and axiomatize there must be mathematical substance. I think that mathematical substance on which we have practiced formalization in the last few decades is near exhaustion and I predict that the next generation will face in mathematics a tough time. ([169], 651)

You should not expect me to describe the mathematical way of thinking much more clearly than one can describe, say, the democratic way of life. ([25], 84)

Although not committing himself to one of the established epistemological or metaphysical doctrines, he (Hilbert) was a philosopher in that he was concerned with the life of the idea as it realizes itself among men and as an indivisible whole; he had the force to evoke it, he felt responsible for it in his own sphere, and measured his individual scientific efforts against it.... No mathematician of equal stature has risen from our generation. ([25], 84)

"Mathematizing" may well be a creative activity of man, like language or music, of primary originality, whose historical decisions defy complete objective rationalization. ([25], 84)

We are not very pleased when we are forced to accept a mathematical truth by virtue of a complicated chain of formal conclusions and computations, which we traverse blindly, link by link, feeling our way by touch. We want first an overview of the aim and of the road; we want to understand the *idea* of the proof, the deeper context. ([169], 646)

Alfred North Whitehead (1861–1947)

I would not go so far as to say that to construct a history of thought without a profound study of the mathematical ideas of successive epochs is like omitting Hamlet from the play which is named after him. That would be claiming too much. But it is certainly analogous to cutting out the part of Ophelia. This simile is singularly exact. For Ophelia is quite essential to the play, she is very charming—and a little mad. Let us grant that the pursuit of mathematics is a divine madness of the human spirit, a refuge from the goading urgency of contingent happenings. ([174], 22)

There is an old epigram which assigns the empire of the sea to the English, of the land to the French, and of the clouds to the Germans. Surely it was from the clouds that the Germans fetched + and –; the ideas which these symbols have generated are much too important to the welfare of humanity to have come from the sea or from the land. ([172], 86)

The science of Pure Mathematics, in its modern developments, may claim to be the most original creation of the human spirit. ([174], 20)

To a missing member of a family group of terms in an algebraical formula.

> Lone and discarded one! divorced by fate,
> Far from thy wished-for fellows—whither art flown?
> Where lingerest thou in thy bereaved estate,
> Like some lost star, or buried meteor stone?
> Thou mindst me much of that presumptuous one
> Who loth, aught less than greatest, to be great,
> From Heaven's immensity fell headlong down
> To live forlorn, self-centered, desolate:
> Or who, like Heraclid, hard exile bore,
> Now buoyed by hope, now stretched on rack of fear,
> Till throned Astæa, wafting to his ear
> Words of dim portent through the Atlantic roar,
> Bade him "the sanctuary of the Muse revere
> And strew with flame the dust of Isis' shore." ([173], 228)

The progress of science consists in observing interconnexions and in showing with a patient ingenuity that the events of this ever-shifting world are but examples of a few general relations, called laws. To see what is general in what is particular, and what is permanent in what is transitory, is the aim of scientific thought. ([172], 4)

It is natural to think that an abstract science cannot be of much importance in the affairs of human life, because it has omitted from its consideration everything of real interest. It will be remembered that Swift, in his description of Gulliver's voyage to Laputa, is of two minds on this point. He describes the mathematicians of that country as silly and useless dreamers, whose attention has to be awakened by flappers. Also, the mathematical tailor measures his height by a quadrant, and deduces his other dimensions by a rule and compasses, producing a suit of very ill-fitting clothes. On the other hand, the mathematicians of Laputa, by their marvellous invention of the magnetic island floating in the air, ruled the country and maintained their ascendency over their subjects. Swift, indeed, lived at a time peculiarly unsuited for gibes at contemporary mathematicians. Newton's *Principia* had just been written, one of the great forces which have transformed the modern world. Swift might just as well have laughed at an earthquake. ([172], 9–10)

The point of mathematics is that in it we have always got rid of the particular instance, and even of any particular sorts of entities. So that for example, no mathematical truths apply merely to fish, or merely to stones, or merely to colours. So long as you are dealing with pure mathematics, you are in the realm of complete and absolute abstraction.... Mathematics is thought moving in the sphere of complete abstraction from any particular instance of what it is talking about. ([174], 22)

Now in creative thought common sense is a bad master. Its sole criterion for judgment is that the new ideas shall look like the old ones. In other words it can only act by suppressing originality. ([172], 116)

What Bacon omitted was the play of a free imagination, controlled by the requirements of coherence and logic. The true method of discovery is like the flight of an aeroplane. It starts from the ground of particular observation; it makes a flight in the thin air of imaginative generalization; and it again lands for renewed observation rendered acute by rational interpretation. ([171], 193)

In the study of ideas, it is necessary to remember that insistence on hard-headed clarity issues from sentimental feeling, as it were a mist, cloaking the perplexities of fact. Insistence on clarity at all costs is based on sheer superstition as to the mode in which human intelligence functions. Our reasonings grasp at straws for premises and float on gossamers for deductions. ([170], 91)

Norbert Wiener (1894–1964)

Mathematics is too arduous and uninviting a field to appeal to those to whom it does not give great rewards. These rewards are of exactly the same character as those of the artist. To see a difficult uncompromising material take living shape and meaning is to be Pygmalion, whether the material is stone or hard, stonelike logic. To see meaning and understanding come where there has been no meaning and no understanding is to share the work of a demiurge. No amount of technical correctness and no amount of labour can replace this creative moment, whether in the life of a mathematician or of a painter or musician. Bound up with it is a judgement of values, quite parallel to the judgement of values that belongs to the painter or the musician. Neither the artist nor the mathematician may be able to tell you what constitutes the difference between a significant piece of work and an inflated trifle; but if he is not able to recognise this in his own heart, he is no artist and no mathematician. ([175], 212)

The Advantage is that mathematics is a field in which one's blunders tend to show very clearly and can be corrected or erased with a stroke of the pencil. It is a field which has often been compared with chess, but differs from the latter in that it is only one's best moments that count and not one's worst. A single inattention may lose a chess game, whereas a single successful approach to a problem, among many which have been relegated to the wastebasket, will make a mathematician's reputation. ([175], 21)

Granted an urge to create, one creates with what one has. With me, the particular assets that I have found useful are a memory of a rather wide scope and great permanence and a free-flowing, kaleidoscope-like train of imagination which more or less by itself gives me a consecutive view of the possibilities of a fairly complicated intellectual situation. The great strain on the memory in mathematical work is not so much the retention of a vast mass of fact in the literature as of the simultaneous aspects of the particular problem on which I have been working and of the conversion of my fleeting impressions into something permanent enough to have a place in memory. For I have found that if I have been able to cram all my past ideas of what the problem really involves into a single comprehensive impression, the problem is more than half solved. What remains to be done is very often the casting aside of those aspects of the group of ideas that are not germane to the solution of the problem. This rejection of the irrelevant and purification of the relevant I can do best at moments in which I have a minimum of outside impressions. Very often these moments seem to arise on waking up; but probably this really means that sometime during the night I have undergone the process of deconfusion which is necessary to establish my ideas. I am quite certain that at least a part of this process can take place during what one would ordinarily describe as sleep, and in the form of a dream. ([175], 212)

Andrew Wiles (1953–)

Perhaps I could best describe my experience of doing mathematics in terms of entering a dark mansion. One goes into the first room, and it's dark, completely dark. One stumbles around bumping into the furniture, and gradually, you learn where each piece of furniture is, and finally, after six months or so, you find the light switch. You turn it on, and suddenly, it's all illuminated. You can see exactly where you were. At the beginning of September, I was sitting here at this desk, when suddenly, totally unexpectedly, I had this incredible revelation. It was the most important moment of my working life. [176]

I never use a computer. I sometimes might scribble. I do doodles. I start trying to find patterns, really, so I'm doing calculations which try to explain some little piece of mathematics, and I'm trying to fit it in with some previous broad conceptual understanding of some branch of mathematics. Sometimes, that'll involve going and looking up in a book to see how it's done there. Sometimes, it's a question of modifying things a bit, sometimes, doing a little extra calculation. And sometimes, you realize that nothing that's ever been done before is any use at all, and you just have to find something completely new. And it's a mystery where it comes from. [176]

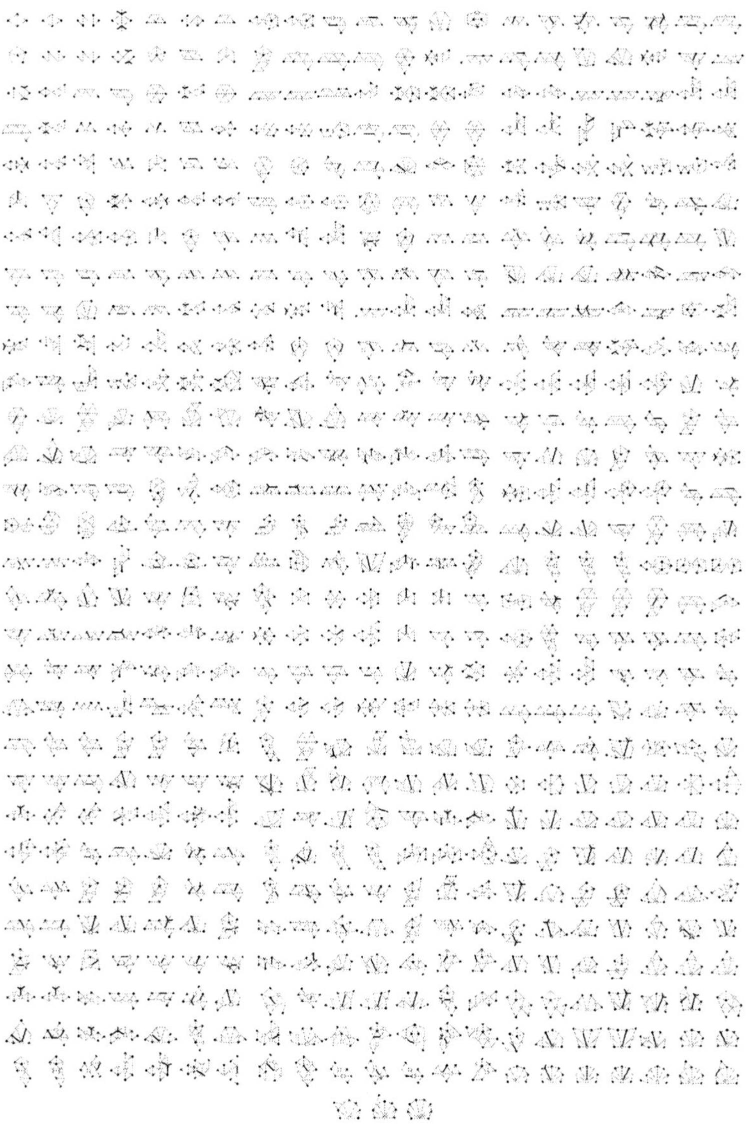

Figure 8. Appel and Haken's computer proof checked the 633 cases shown above.

$1.37\ldots^{\pi} = e$ $\quad e\sqrt{1+e\sqrt{1+e\sqrt{1+e\sqrt{1+e\sqrt{\ldots}}}}} = 1.37\ldots$ $\quad \frac{2.2222\ldots}{\phi} = 1.37\ldots$

$\frac{19^2}{\phi^2} = 137\ldots$ $\quad \frac{\phi^2}{19} = .137\ldots$ $\quad 1.37\ldots^{e} - 1.37\ldots = 1.$ $\quad \frac{12.3456789}{9} = 1.37\ldots$

$(1.9)^{\frac{1}{2}} = 1.37\ldots$

$\pi^{-1.37\phi^2} = 1.37\ldots$ $\quad \frac{\sqrt{e+\sqrt{e+\sqrt{e+\ldots}}}}{\phi} = 1.37\ldots$ $\quad \frac{(111111)^2}{9} = 1.37\ldots\times 10^9$

$(111111)^3 = 1.37\ldots\times 10^{15}$

$3.71^2 = 13.7\ldots$ $\quad (2\phi)^{1.37\ldots} = 5.$ $\quad \frac{\sqrt[3]{11}}{\phi} = 1.37\ldots$ $\quad (1.1)^{\sqrt{11}} = 1.37\ldots$

$\phi^{3.7^{-\frac{1}{\pi}}} = 1.37\ldots$ $\quad \frac{\pi}{\sqrt{2}\,\phi} = 1.37\ldots$ $\quad (\phi^2+1)^{\frac{1}{4}} = 1.37\ldots$ $\quad \phi^{1.111^{-4}} = 1.37\ldots$

$37! = 1.37\ldots\times 10^{43}$ $\quad \phi^{\ln\phi(1.37)} = 1.37\ldots$

$.037037\times 37 = 1.37\ldots$ $\quad (\pi-\phi)^{\frac{3}{4}} = 1.37\ldots$ $\quad \frac{1.37\ldots}{\phi^2} = 1.37\ldots^{-2}$ $\quad (1.37\phi^2)^{\frac{1}{4}} = 1.37\ldots$

$1.37\ldots^{-2\pi} = .137\ldots$ $\quad \left(\frac{1.37\sqrt{5}}{\phi}\right)^{\frac{1}{2}} = 1.37\ldots$ $\quad \frac{1.37}{\phi} = 1.37\ldots^{-\frac{1}{2}}$ $\quad (\pi\phi)^{\frac{1}{\pi\phi}} = 1.37\ldots$

$\frac{\pi^{-\pi}}{.02} = 1.37\ldots$ $\quad \phi^{\frac{2}{3}} = 1.37\ldots$ $\quad 1.37\ldots^{\pi} = e^{\frac{137}{37}}$ $\quad = (13.7)^{\frac{1}{2}}$ $\quad \pi^{.28} = 1.37\ldots$

$\frac{137.0137}{73} = (1.37\ldots)^2$ $\quad \frac{6}{\phi} = (13.7\ldots)^{\frac{1}{2}}$

$\frac{(1.37\ldots)^{\pi}}{2} = 1.37\ldots$ $\quad 1.37\ldots^{\phi} = \frac{5}{3}$ $\quad \pi^{\frac{1}{1.37\phi^2}} = 1.37\ldots$ $\quad \ln\left(\frac{137^3}{\phi^2}\right) = 13.7\ldots$

$\log_{10} e^{3.17} = 1.37\ldots$ $\quad \frac{\phi^2}{\sqrt{5}} = (1.37\ldots)^{\frac{1}{2}}$ $\quad \frac{40}{\phi^7} = 1.37\ldots$ $\quad \frac{\phi^4}{5} = 1.37\ldots$

$\phi^{\left(\frac{\phi}{2}\right)^2} = 1.37\ldots$ $\quad \frac{\phi^3}{\sqrt{5}} = (1.37\ldots)^2$ $\quad e^{\frac{1}{\pi}} = 1.37\ldots$ $\quad e^{\frac{1}{3.17}} = 1.37\ldots$ $\quad 2\phi^4 = 13.7\ldots$

$\frac{\phi^5}{\sqrt{5}} = (1.37\ldots)^5$ $\quad e^{\frac{1}{\sqrt{10}}} = 1.37\ldots$ $\quad e^{.317} = 1.37\ldots$

$\phi^{\ln\frac{\pi}{\phi}} = 1.37\ldots$ $\quad e^{\phi^2} = 13.7\ldots$ $\quad \log_{10} e^{\sqrt{10}} = 1.37\ldots$

$\frac{37}{27} = 1.37\ldots$ $\quad \phi^{1.37\ldots} = \frac{\pi}{\phi}$; $\log_{10} 1.37\ldots = .137\ldots$

$\log_{\phi}\frac{\pi}{\phi} = 1.37\ldots$ $\quad \left(\frac{20}{9\phi}\right)^{1.37\ldots}$ $\quad \phi^{1.37\ln\phi} = 1.37\ldots$ $\quad \left(\frac{2\pi}{\phi^2}\right)^{\frac{1}{e}} = 1.37\ldots$ $\quad \log_{10}(2\phi e^2) = 1.37\ldots$

$2^{13.7\ldots} = 1.37\ldots\times 10^4$

$\phi^{\pi^{-\frac{1}{e}}} = 1.37\ldots$ $\quad \pi^{1.37\ldots^{-4}} = 1.37\ldots$ $\quad 2^{7.1\ldots} = 137.$ $\quad 2^{37} = 1.37\ldots\times 10^{11}$

$\left(\frac{\pi^5}{\phi}\right)^{\frac{1}{2}} = 13.7\ldots$ $\quad \frac{\ln 37}{\phi^2} = 1.37\ldots$ $\quad 36^{1.37\ldots} = 137.$

$\pi^{1.5^{-\pi}} = 1.37\ldots$ $\quad \frac{3.6}{\phi^2} = 1.37\ldots$ $\quad 3.6^{\frac{1}{4}} = 1.37\ldots$

$163^{\frac{1}{16}} = 1.37\ldots$ $\quad \frac{\phi}{\pi} + 1.37 = (1.37)^2$ $\quad \frac{13.7^{1.37}}{\phi^2} = 13.7\ldots$ $\quad \frac{1}{3^6} = 1.37\ldots\times 10^{-3}$

$\phi^{\frac{\phi^2}{4}} = 1.37\ldots$ $\quad \frac{20}{9\phi} = 1.37\ldots$ $\quad \frac{7}{\pi\phi} = 1.37\ldots$ $\quad (\ln 13.7)^{\frac{1}{3}} = 1.37\ldots$ $\quad \frac{100^{\frac{1}{\pi}}}{\pi} = 1.37\ldots$

$e^{\phi^2} = 13.7\ldots$ $\quad \frac{\pi\phi}{37} = .137\ldots$ $\quad (e\pi)^{-2} = .0137\ldots$ $\quad \left(\frac{13.7}{2}\right)^{\frac{1}{6}} = 1.37\ldots$

$(73\pi\phi)^2 = 1.37\ldots\times 10^5$ $\quad \phi^{\frac{2}{3}} = 1.37\ldots$ $\quad \phi^{2^{(\pi-\phi)}} = (1.37\ldots)^{\pi}$ $\quad 5^{\frac{1}{5}} = 1.37\ldots$

$\frac{17\phi}{2} = 13.7\ldots$ $\quad 9(\pi-\phi) = 13.7\ldots$ $\quad (3.71)^2 = 13.7\ldots$ $\quad \frac{10\phi}{e\pi} = (1.37\ldots)^2$ $\quad 18^{\frac{1}{9}} = 1.37\ldots$

$(\pi-\phi)^{(\pi-\phi)} = (1.37\ldots)^2$ $\quad \phi^{\frac{1}{\pi-\phi}} = 1.37\ldots$ $\quad \phi^{\left(\frac{\phi^2}{4}\right)} = 1.37\ldots$ $\quad 13^{\frac{1}{8}} = 1.37\ldots$

Figure 9. These formulas come from a large list of over 200 mathematical coincidences involving 137, compiled by Gary Adamson. ([121], 270–271)

Appendix A:
Hadamard's Survey

1. At what time, as well as you can remember, and under what circumstances, did you begin to be interested in mathematical sciences? Have you inherited your liking for mathematical sciences? Were any of your immediate ancestors or members of your family (brothers, sisters, uncles, cousins, etc.) particularly good at mathematics? Was their influence or example to any extent responsible for your propensity for mathematics?[1]

2. Toward what branches of mathematical science did you feel especially attracted?

3. Are you more interested in mathematical science per se or in its application to natural phenomena?

4. Have you a distinct recollection of your manner of working while you were pursuing your studies, when the goal was rather to assimilate the results of others than to indulge in personal research? Have you any interesting information to offer on that point?

5. After having completed the regular course of mathematical studies (which, for instance, corresponds to the program of the Licence mathematique or of two Licences or of the Aggregation) in what direction did you consider it expedient to continue your studies? Did you endeavor, in the first place, to obtain a general and extensive knowledge of several parts of science before writing or publishing anything of consequence? Did you, on the contrary, at first try to penetrate rather deeply into a special subject, studying almost exclusively what was strictly requisite for that purpose, and only afterwards extending your studies little by little? If you have used other methods, can you indicate them briefly? Which one do you prefer?

6. Among the truths which you have discovered, have you attempted to determine the genesis of those you consider the most valuable?

---

1 Based on ([51], 137–141). This version of Hadamard's survey has been edited to update the language and form of the original early 20th century translation.

7. What, in your estimate, is the role played by chance or inspiration in mathematical discoveries? Is this role always as great as it appears to be?

8. Have you noticed that, occasionally, discoveries or solutions on a subject entirely foreign to the one you are dealing with occur to you and that these relate to previous unsuccessful research efforts of yours?

9. Would you say that your principal discoveries have been the result of deliberate endeavor in a definite direction, or have they arisen, so to speak, in your mind?

10. Have you ever worked in your sleep or have you found in dreams the answers to problems? Or, when you waken in the morning, do solutions which you had vainly sought the night before, or even days before, or quite unexpected discoveries, present themselves ready-made to your mind?

11. When you have arrived at a conclusion about something you are investigating with a view to the publication of your findings, do you immediately write down the part of your work to which your discovery applies; or do you let your conclusions accumulate in the form of notes and begin the redaction of the work only when its contents are important enough?

12. Generally speaking, how much importance do you attach to reading for mathematical research? What advice in this respect would you give to a young mathematician who has had the usual classical education?

13. Before beginning a piece of research work, do you first attempt to assimilate what has already been written on that subject?

14. Or do you prefer to leave your mind free to work unbiased and do you only afterwards verify by reading about the subject so as to ascertain just what is your personal contribution to the conclusions reached?

15. As far as method is concerned, do you make any distinction between invention and redacting?

16. Does it seem to you that your habits of work are appreciably the same as they were before you had completed your studies?

17. When you take up a question, do you try to make as general a study as possible of the more or less specific problems which occur to you? Do you usually prefer, first to study special cases or a more inclusive one, and then to generalize progressively?

18. In your principal research studies, have you followed the same line of thought steadily and uninterruptedly to the end, or have you laid it aside at times and subsequently taken it up again?

19. What is, in your opinion, the minimum number of hours during the day, week, or the year, which a mathematician who has other demands on his time should devote to mathematics so as to study profitably certain branches of these same mathematics? Do you believe that one should, if one can, study a little every day, say for one hour at least?

20. Do artistic and literary occupations, especially those of music and poetry, seem to you likely to hamper mathematical invention, or do you think they help it by giving the mind temporary rest?

21. What are your favorite hobbies, pursuits, or chief interests, aside from mathematics, in your leisure time? Do metaphysical, ethical, or religious questions attract or repel you?

22. If you are absorbed by professional duties, how do you fit these into your personal studies?

23. What council, in brief, would you offer to a young man studying mathematics? To a young mathematician who has finished the usual course of study and desires to follow a scientific career?

Questions about daily habits

24. Do you believe that it is beneficial to a mathematician to observe a few special rules of hygiene such as diet, regular meals, time for rest, etc.?

25. What do you consider the normal amount of sleep necessary?

26. Would you say that a mathematician's work should be interrupted by other occupations or by physical exercises which are suited to the individual's age and strength?

27. Or, on the contrary, do you think one should devote the whole day to one's work and not allow anything to interfere with it; and, when it is finished, take several days of complete rest? Do you experience definite periods of inspiration and enthusiasm succeeded by periods of depression and incapacity to work? Have you noticed whether these intervals alternate regularly and, if so, how many days, approximately, does the period of activity last and also the period of inertia? Do physical or meteorological conditions (i.e. temperature, light, darkness, the season of the year, etc.) exert an appreciable influence on your ability to work?

28. What physical exercises do you do, or have you done, as relaxation form mental work? Which do you prefer?

29. Would you rather work in the morning or in the evening?

30. If you take a vacation, do you spend it studying mathematics (if so, to what extent?) or do you devote the entire time to rest and relaxation?

31. Does one work better standing, seated or lying down? Does one work better standing at the blackboard or on paper? To what extent is one disturbed by outside noises? Can one pursue a problem while walking or in a train? How do stimulants or sedatives (tobacco, coffee, alcohol, etc.) affect the quality and quantity of ones work?

32. It would be very helpful for the purpose of psychological investigation to know what internal or mental images, what kind of "internal words, mathematicians make use of; whether they are motor, auditory, visual, or mixed, depending on the subject which they are studying.

33. Especially in research thought, do the mental pictures or internal words present themselves in the full consciousness or in the fringe-consciousness? The same question is asked concerning the arguments which these mental pictures or words may symbolize.

Appendix B:
Biographies

Andrews, George Eyre

George Eyre Andrews was born December 4, 1938 in Salem, Oregon. He is currently the Evan Pugh Professor of Mathematics at Pennsylvania State University. In 1964, he received his Ph.D. at University of Pennsylvania under Hans Rademacher. His research centers on the theory of partitions and related areas. Andrews has a long-term interest in the work of Ramanujan and in 1976, he unearthed his lost notebook in the Trinity College Library at Cambridge.

Askey, Richard Allen

Richard Allen Askey was born June 4, 1933 in St. Louis, Missouri. He received his Ph.D. from Princeton University in 1961, where his advisor was Salomon Bochner. He is considered the world's foremost authority on Special Functions.

Atiyah, Michael Francis

Michael Francis Atiyah was born April 22, 1929 in London, England. He obtained his Ph.D. at Cambridge in 1955, where his advisor was W.V.D. Hodge. He has contributed to a wide range of topics in mathematics centring around the interaction between geometry and analysis. In 1966, he was awarded the Fields Medal for his work in developing K-theory. Atiyah was knighted in 1983 and made a member of the Order of Merit in 1992.

Bellman, Richard Ernest

Richard Ernest Bellman was born August 26, 1920 in Brooklyn, New York City. He received his Ph.D. in 1947 from Princeton University, where his advisor was Solomon Lefschetz. For his outstanding contributions to modern control theory and systems analysis as well as his invention of dynamic programming, he was given the John von Neumann Theory Award, and the IEEE Medal of Honour, amongst others. Professor Bellman died on March 19, 1984.

Berlekamp, Elwyn Ralph

Elwyn Ralph Berlekamp was born September 6, 1940 in Dover, Ohio. In 1964, he received his Ph.D. in electrical engineering at MIT, where his advisor was Robert Gallager. He invented the Berlekamp-Massey algorithm used to implement Reed-Solomon error correction. He is also recognized as one of the founders of combinatorial game theory, and is now a professor of mathematics at the University of California, Berkeley.

Bers, Lipman

Lipman Bers was born May 22, 1914 in Riga, Latvia. In 1938, he received his Ph.D. from the Charles University of Prague under the supervision of Karl Loewner. Not only has he made important contributions to mathematics of Riemann surfaces, Kleinian groups and Teichmüller theory, he was also committed to the protection of human rights throughout his life. Professor Bers died on October 29, 1993.

Blackwell, David Harold

David Harold Blackwell was born April 24, 1919 in Centralia, Illinois. He received his doctorate in 1941 from the University of Illinois-Urbana Champaign under the supervision of Joseph Doob. For his work in statistics and game theory, he was elected to the National Academy of Sciences (1965) and received the John von Neumann Prize (1979).

Boas Jr., Ralph Phillip

Ralph Philip Boas, Jr. was born August 8, 1913 in Walla Walla, Washington. He received his Ph.D. in 1937 at Harvard University under the supervision of David Widder. His extensive contributions to real and complex analysis have earned him many honours including a Guggenheim Fellowship (1951–52) and the Distinguished Service Award from the Mathematical Association of America in 1981. Professor Boas died on July 25, 1992.

Bombieri, Enrico

Enrico Bombieri was born November 26, 1940 in Milan, Italy. He received his Ph.D. from Universitá di Milano in 1963, where his advisor was Giovanni Ricci. One of the world's leading authorities on number theory and analysis, he was awarded the Fields Medal in 1974 for his work on the large sieve and its application to the distribution of prime numbers. He is currently at the Institute for Advanced Study.

Christodoulou, Demetrios

Demetrios Christodoulou was born October 19, 1951 in Athens, Greece. He received his Ph.D. in physics from Princeton University in 1971 under the supervision of John Archibald Wheeler. He is a recipient of the Bôcher Memorial Prize for his contributions to the mathematical theory of general relativity. Since 1992, he has been a professor of mathematics at Princeton University.

Cohen, Paul Joseph

Paul Joseph Cohen was born April 2, 1934 in Long Branch, New Jersey. He received his Ph.D. in 1958 from the University of Chicago under the supervision of Antoni Zygmund. He won two of the most prestigious awards in mathematics—in completely different fields. In 1964, he won the Bôcher Memorial Prize for analysis and in 1966, he was awarded a Fields Medal for his fundamental work on the foundations of set theory, which solved Hilbert's continuum hypothesis problem. Professor Cohen died on March 23, 2007.

Conway, John Horton

John Horton Conway was born December 26, 1937 in Liverpool, England. He received his doctorate in 1967 from Cambridge under the supervision of Harold Davenport. For his extensive contributions to fields such as combinatorial game theory, geometry, group theory, and number theory, he has received the Berwick Prize (1971), the Pólya Prize (1987), and the Nemmers Prize in Mathematics (1998).

Dantzig, George Bernard

George Bernard Dantzig was born November 8, 1914 in Portland, Oregon. He received his Ph.D. in 1946 from the University of California, Berkeley. He is recognized for inventing the simplex algorithm and is considered the founder of linear programming. For his outstanding work, he has received the National Medal of Science in 1975 and the John von Neumann Prize in 1974. Professor Dantzig died on May 13, 2005.

de Boor, Carl R.

Carl R. de Boor was born December 3, 1937 in Stolp, Germany. He received his Ph.D. from the University of Michigan in 1966. He has made contributions to numerical analysis and approximation theory. In 1996, he won the John von Neumann Prize, and he later received the 2003 National Medal of Science in mathematics.

Deligne, Pierre René

Pierre René Deligne was born October 3, 1944 in Etterbeek, Brussels, Belgium. He was awarded his Ph.D. by the Free University of Brussels in 1968, where his advisor was Alexandre Grothendieck. In 1978, he was awarded the Fields Medal for his work in algebraic topology, and he received the 2004 Balzan Prize in Mathematics.

Diaconis, Persi Warren

Persi Warren Diaconis was born January 31, 1945 in New York City. He was awarded his Ph.D. in mathematical statistics from Harvard University in 1974. He has made pioneering contributions to probability and statistics, and has received numerous honours including election to membership in the National Academy of Sciences (1995) and the American Philosophical Society (2005). He was also President of The Institute of Mathematical Sciences (1997–1998).

Dirac, Paul Adrien Maurice

Paul Adrien Maurice Dirac was born August 8, 1902 in Bristol, England. He received his Ph.D. in 1926 from the University of Cambridge, where his advisor was Ralph Fowler. For his work on antiparticles and wave mechanics, he received the Nobel Prize in physics in 1933. The previous year he had been appointed the Lucasian Professor of Mathematics at Cambridge and held the position for 37 years. Professor Dirac died on October 20, 1984.

Donoho, David Levin

David Levin Donoho was born September 13, 1965 in Redwood City, California. He received his Ph.D. in statistics at Harvard University where his thesis advisor was Peter J. Huber. He has done ground-breaking work in statistical theory and is currently at Stanford University.

Doob, Joseph Leo

Joseph Leo Doob was born February 27, 1910 in Cincinnati, Ohio. He obtained his Ph.D. in 1932 from Harvard University, where his advisor was J.L. Walsh. He was a pioneer in the field of probability and measure theory. For his contributions, he was awarded the National Medal of Science in 1979, and in 1984, given the Steele Prize. Professor Doob died on June 7, 2004.

Dyson, Freeman John

Freeman John Dyson was born December 15, 1923 in Crowthorne, England. He is famous for his unification of the three versions of quantum electrodynamics invented by Feynman, Schwinger and Tomonaga. He has also worked on nuclear reactors, solid-state physics, astrophysics, and biology and for his contributions, received numerous awards including the Lorentz Medal, the Hughes Medal, the Max Planck Medal, and the Enrico Fermi Award.

Efron, Bradley

Bradley Efron was born May 1938 in St. Paul, Minnesota. He received his Ph.D. in 1964 from Stanford University under the supervision of Rupert Miller, Jr. He is best known for proposing the bootstrap resampling technique.

Erdős, Paul

Paul Erdős was born March 26, 1913 in Budapest, Hungary. In 1934, he received his Ph.D. from University Pázmány Péter, where his advisor was Leopold Fejér. He is considered one of the greatest and most prolific mathematicians of the 20th century, who posed and solved innumerable problems in number theory, combinatorics, geometry, and many other branches of mathematics. Professor Erdős died on September 20, 1996.

Faltings, Gerd

Gerd Faltings was born July 28, 1954 in Gelsenkirchen-Buer, Germany. He was awarded his Ph.D. in 1978 from the University of Münster under the supervision of Hans-Joachim Nastold. In 1986, he was awarded a Fields Medal for his proof of the Mordell Conjecture using methods of arithmetic algebraic geometry. He is also linked with the work leading to the final proof of Fermat's Last Theorem.

Feferman, Solomon

Solomon Feferman was born December 13, 1928 in New York, New York. He was awarded his Ph.D. in 1957 from the University of California, Berkeley, where his advisor was Alfred Tarski. He received the Rolf Schock Prize in Logic and Philosophy for 2003 for his work in mathematical logic.

Fefferman, Charles Louis

Charles Louis Fefferman was born April 18, 1949 in Silver Spring, Maryland. He received his Ph.D. in 1969 from Princeton University under the supervision of Elias Stein. In 1978, he was awarded the Fields Medal for his work on partial differential equations and Fourier analysis. At 22, he was the youngest full professor ever appointed in the United States. He currently is at Princeton working on such diverse fields as mathematical physics, neural networks, and fluid dynamics.

Fleming, Wendell H.

Wendell H. Fleming was born March 7, 1928 in Guthrie, Oklahoma. Fleming received a Ph.D. in 1951 from the University of Wisconsin. For his work in calculus of variations, geometric measure theory, mathematical finance, and stochastic control, amongst others, he was awarded a Steele Prize from the American Mathematical Society and a Reid Prize from the Society of Industrial and Applied Mathematics.

Gårding, Lars

Lars Gårding, a Swedish mathematician, was born in 1919. He has made considerable contributions to the study of partial differential operators. He is a professor emeritus of mathematics at Lunds Universitet, Sweden.

Giorgi, Ennio de

Ennio de Giorgi was born February 8, 1928 in Lecce, Italy. He received his *laurea*, the diploma which marks the end of the Italian undergraduate curriculum, in Rome in 1950. He has made important contributions to many fields including geometric measure theory and partial differential equations, as well as providing the crucial step to solving Hilbert's nineteenth problem. In 1959, he was appointed full professor at the Scuola Normale Superiore di Pisa where he worked and taught for almost forty years. Professor Giorgi died on October 25, 1996.

Gleason, Andrew Mattei

Andrew Mattei Gleason was born November 4, 1921 in Fresno, California. Appointed as a Junior Fellow, he was one of the few professors at Harvard University who have never obtained a Ph.D. He was best known for his work on Hilbert's Fifth Problem. He received many honours for his contributions to teaching, research, and mathematics in general. During World War II, he played a role in breaking Japanese codes. Professor Gleason died on October 18, 2008.

Gödel, Kurt Friedrich

Kurt Friedrich Gödel was born April 28, 1906 in Brünn, Austria-Hungary. In 1929, he obtained his doctorate from the University of Vienna under Hans Hahn. Gödel is best known for his proof of "Gödel's Incompleteness Theorems" and contributed fundamental results about axiomatic systems. From 1953 until his death, he held a chair at Princeton, where he was a close friend of Einstein's. Professor Gödel died on January 25, 1978.

Graham, Ronald Lewis

Ronald Lewis Graham was born October 31, 1935 in Taft, California. In 1962, he obtained his Ph.D. from the University of California, Berkeley, where his advisor was D. H. Lehmer. He has made outstanding contributions to scheduling theory, computational geometry, Ramsey theory, and quasi-randomness. In a 1977 paper of his, the largest number ever used in a mathematical proof became known as Graham's number. He had a close relationship with Paul Erdős. In 2003, he won the Steele Prize for Lifetime Achievement.

Grenander, Ulf

Ulf Grenander was born July 23, 1923 in Vastervik, Sweden. He received his Ph.D. from Stockholm University in 1950, where his advisor was Harald Cramér. He is well known for ground-breaking research in pattern theory, computer vision and actuarial mathematics, amongst others.

Grothendieck, Alexandre

Alexandre Grothendieck was born March 28, 1928 in Berlin, Germany. He received his Ph.D. from the University of Nancy, where his advisor was Laurent Schwartz. He has made major contributions to algebraic geometry, homological algebra, and functional analysis. In 1966, he was awarded a Fields Medal for his outstanding work. In contrast, he declined the Crafoord Prize, co-awarded with Pierre Deligne in 1988.

Guy, Richard Kenneth

Richard Kenneth Guy was born September 30, 1916 in Nuneaton, Warwickshire. He received his masters in mathematics from Cambridge in 1941. He has published over 100 papers and books covering combinatorial game theory, number theory, and graph theory.

Halmos, Paul

Paul Halmos was born March 3, 1916 in Budapest, Hungary. He obtained a Ph.D. in 1938 from the University of Illinois under the supervision of Joseph Doob. He is known for his outstanding contributions to operator theory, ergodic theory, and functional analysis. He has received many honours for his work, including the Steele Prize in 1983. Professor Halmos died on October 2, 2006.

Harish-Chandra

Harish-Chandra was born October 11, 1923 in Kanpur, India. He received his Ph.D. in 1947 from Cambridge, where his advisor was Paul Dirac. For his fundamental work in representation theory of reductive groups, he received many honours including fellowship in the Royal Society of London in 1973. Harish-Chandra died on October 16, 1983.

Hochster, Melvin

Melvin Hochster was born August 2, 1943 and received his Ph.D. from Princeton University in 1967, where his advisor was Goro Shimura. He is best known for his work in commutative algebra and has received many honours for his contributions to research and teaching. He is currently the Jack E. McLaughlin Distinguished University Professor of Mathematics at the University of Michigan.

Huber, Peter

Peter Jost Huber was born March 25, 1934 in Wohlen, Switzerland. He received his Ph.D. from ETH Zürich in 1961 under the supervision of Beno Eckmann and Heinz Hopf. He is most known for his fundamental work in robust statistics, but has also made important contributions to computational statistics, strategies in data analysis, and applications of statistics in fields such as crystallography, EEGs, and human growth curves. He has also written on Babylonian mathematics, astronomy and history.

Ivrii, Victor

Victor Ivrii was born January 10, 1949 in Sovetsk, USSR. He received his Ph.D. from Novosibirsk State University under the supervision of S.L. Sobolev in 1973. He is currently working on research in analysis, spectral theory, and partial differential equations at the University of Toronto.

Kac, Marc

Mark Kac was born August 3, 1914 in Krzemieniec, Poland. In 1937, he was awarded his doctorate from the University of Lvov. His main contributions were made in applications of probability to statistical physics, and many will remember his paper *Can One Hear the Shape of a Drum?* for which he received the Chauvenet Prize from the MAA in 1968. Professor Kac died on October 26, 1984.

Kaplansky, Irving

Irving Kaplansky was born March 22, 1917 in Toronto, Ontario. He received his Ph.D. from Harvard University in 1941, where his advisor was Saunders Mac Lane. He has made major contributions to ring theory, group theory, and field theory. For his outstanding work, he was awarded a Guggenheim Fellowship in 1948, and the Steele Prize in 1989.

Kleitman, Daniel J.

Daniel J. Kleitman was born in New York, NY, on October 4, 1934. He received his Ph.D. from Harvard University in 1958 under the supervision of Julian Schwinger and Roy Glauber. He is currently a professor of applied mathematics at MIT and his main research interests include combinatorics and graph theory.

Kline, Morris

Morris Kline was born May 1, 1908 in Brooklyn, New York. He earned his doctorate in 1936 from New York University and went on to write many papers and more than a dozen books on the history, philosophy, and teaching of mathematics.

Knuth, Donald Ervin

Donald Ervin Knuth was born January 10, 1938 in Milwaukee, Wisconsin. He received his Ph.D. from the California Institute of Technology in 1963. For his outstanding contributions to computer science, he has received the Turing Award, the National Medal of Science, the John von Neumann Medal and the Kyoto Prize. He is currently a Professor Emeritus of the Art of Computer Programming at Stanford University.

Lakatos, Imre

Imre Lakatos was born November 9, 1922 in Hungary. He obtained his doctorate in philosophy from the University of Cambridge in 1961 and made significant contributions to the philosophy of mathematics and science. Professor Lakatos died on February 2, 1974.

Landau, Susan

Susan Landau received her Ph.D. from MIT in 1983. From 1999-2010 Landau was a Distinguished Engineer at Sun Microsystems Laboratories, where she worked on security, cryptography, and policy. She works mainly in algebraic algorithms and computational complexity. Currently, she is a Distinguished Engineer at Sun Microsystems Laboratories.

Lang, Serge

Serge Lang was born May 19, 1927 near Paris, France. He earned his Ph.D. in 1951 from Princeton University, where his advisor was Emil Artin. He was influential in numerous areas of mathematics including number theory and algebraic geometry. His many recognitions include the AMS Cole Prize (1959), the AMS Steele Prize (1999), as well as membership to the National Academy of Sciences in 1985. Professor Lang died on September 12, 2005.

Lax, Peter David

Peter David Lax was born May 1, 1926 in Budapest, Hungary. He received his Ph.D. in 1949 from New York University, where his advisor was Kurt Friedrichs. He has made important contributions to integrable systems, fluid dynamics, and mathematical computing, amongst others.

For his versatile work, he was awarded the National Medal of Science in 1986, the Wolf Prize in 1987 and the Abel Prize in 2005.

Le Cam, Lucien Marie

Lucien Marie Le Cam was born November 18, 1924 in Croze, Creuse, France. He obtained his Ph.D. in 1952 at the University of California, Berkeley, where his advisor was Jerzy Neyman. He was the key figure in the development of the asymptotic theory of statistics. Professor Le Cam died on April 25, 2000.

Lichnerowicz, André

André Lichnerowicz was born January 21, 1915 in Bourbon l'Archambault, France. He obtained his Ph.D. in 1939 under the direction of Georges Darmois. Not only was he an active researcher, publishing more than 350 articles and books, he made many efforts towards educational reform in France. Professor Lichnerowicz died December 11, 1998.

Malliavin, Paul

Paul Malliavin was born September 10, 1925. He received his Ph.D. under the supervision of Szolem Mandelbrojt. He has contributed to many fields including probabilities calculus, stochastic calculus of variations, harmonic analysis and approximation theory.

Mandelbrot, Benoit

Benoit Mandelbrot was born November 20, 1924 in Warsaw, Poland. He received his Ph.D. from Université de Paris in 1952. He is known as the creator of fractal geometry. For his remarkable achievements, he has been awarded numerous honors including the Wolf Prize for physics in 1993.

Marsden, Jerrold Eldon

Jerrold Eldon Marsden was born August 17, 1942 in Ocean Falls, British Columbia, Canada. He received his Ph.D. in applied mathematics from Princeton University. His contributions to a variety of topics such as Hamiltonian systems, fluid mechanics, plasma physics, and solar system mission design earned him many honours. In 2000, he received the Max Planck Research Award for Mathematics and Computer Science and in

2006, he was named a Fellow of the Royal Society of London. Professor Marsden passed away on September 21, 2010.

Mazur, Barry Charles

Barry Charles Mazur was born December 19, 1937 in New York City. He received his Ph.D. from Princeton University in 1959, and is currently a professor of mathematics at Harvard. He is a member of the National Academy of Sciences and has been awarded the Veblen Prize (in geometry) and the Cole Prize (in number theory) from the AMS.

McDuff, Dusa

Dusa McDuff was born October 18, 1945 in London, England. She obtained her Ph.D. in 1971 from Cambridge University under the supervision of G.A. Reid. She is best known for her work in the geometry of multi-dimensional structures. For her outstanding contributions to mathematics, she was elected a Fellow of the Royal Society of London in 1994.

McKean, Henry P.

Henry P. McKean was born December 14, 1930 in Wenham, Massachusetts. He received his Ph.D. from Princeton University in 1955, where his advisor was Willi Feller. His work in probability, partial differential equations, and complex function theory has earned him many honours including being named a Guggenheim Fellow in 1973. He is currently at New York University.

McMullen, Curtis Tracy

Curtis Tracy McMullen was born May 21, 1958. He received his Ph.D. in 1985 from Harvard University under the supervision of Dennis Sullivan. In 1998, he won the Fields Medal for his work in the fields of geometry and complex dynamics. He is currently a Professor of Mathematics at Harvard University.

Meyer, Yves F.

Yves F. Meyer was born July 19, 1939. He is a pioneer in wavelet analysis. He is currently a professor at the École Normale Supérieure de Cachan.

Morawetz, Cathleen Synge

Cathleen Synge Morawetz was born May 5, 1923 in Toronto, Canada. She earned her Ph.D. at New York University in 1951 where her advisor was Kurt Friedrichs. For her pioneering work in partial differential equations and wave propagation, she was given many honours including a 1998 National Medal of Science.

Neumann, John von

John von Neumann was born December 28, 1903 in Budapest, Hungary. He obtained his Ph.D. in 1926 from Eötvös Loránd University under the direction of Leopold Fejér. One of history's outstanding mathematicians, he made contributions to quantum mechanics, functional analysis, set theory, topology, game theory, and many other mathematical fields. For his work, he received many awards including the Bôcher Prize in 1938, the Medal for Merit in 1947, and the Medal for Freedom in 1956. Professor von Neumann died on February 8, 1957.

Nirenberg, Louis

Louis Nirenberg was born February 28, 1925 in Hamilton, Ontario, Canada. He received his doctorate from New York University in 1949 under the direction of James Stoker. He has received many honours for his work on partial differential equations including the Crafoord Prize, the Bôcher Prize, the Jeffrey-Williams Prize, the Steele Prize, and the National Medal of Science.

Papanicolaou, George

George Papanicolaou was born January 23, 1943 in Athens, Greece. He received his Ph.D. in 1969 from New York University, where his advisor was Joseph Keller. For his work in waves and diffusion in inhomogeneous or random media as well as in analytic and stochastic methods, he has received many honours including a Guggenheim Fellowship in 1983 and the John von Neumann Prize in 2006.

Papert, Samuel

Samuel Papert was born February 29, 1928 in Pretoria, South Africa. He received his first Ph.D. from the University of Witwatersrand in 1952, then another from the University of Cambridge in 1959. He became a research associate at MIT in 1963, and developed constructionism based on the work of Jean Piaget's learning theories. His work has been influential to researchers in the fields of education and computer science.

Penrose, Roger

Sir Roger Penrose was born August 8, 1931 in Colchester, Essex, England. He received his Ph.D. at Cambridge University in 1958 and went on to become the Rouse Ball Professor of Mathematics at Oxford University. He is best known for his outstanding work in mathematical physics, particularly general relativity and cosmology, as well as his later work on non-periodic tilings.

Peskin, Charles Samuel

Charles Samuel Peskin was born June 1947 and obtained his Ph.D. in Physiology in 1972 from Yeshiva University under the supervision of Alexandre Chorin. He had earlier obtained a B.A. in engineering and applied physics at Harvard University in 1968. He is best known for his work in modelling and simulation in medicine and the life sciences, especially using immersed boundary methods to model heart physiology. In 2003, he was awarded the AMS-SIAM Birkhoff Prize in Applied Mathematics.

Rényi, Alfréd

Alfréd Rényi was born March 30, 1921 in Budapest, Hungary. He obtained his Ph.D. in 1947 from the University of Szeged under the supervision of Frigyes Riesz. Best known for his work in probability and number theory, he was the founder of the Mathematical Institute of the Hungarian Academy of Sciences. Professor Rényi died on February 1, 1970.

Ribet, Kenneth Alan

Kenneth Alan Ribet was born June 28, 1948 in New York. He received his Ph.D. in 1973 from Harvard University, where his advisor was John Tate. He is known for his work in number theory and algebraic geometry. Among the honours he has received include the 1989 Fermat Prize, for his work in number theory and Fermat's Last Theorem.

Robinson, Julia Bowman

Julia Bowman Robinson was born December 8, 1919 in St. Louis, Missouri. In 1948, she received her Ph.D. from the University of California at Berkeley under Alfred Tarski. She is best known for her work in decision problems and Hilbert's Tenth Problem. During 1983–84 she served as President of the American Mathematical Society, the first woman to hold that position. Professor Robinson died on July 30, 1985.

Rota, Gian-Carlo

Gian-Carlo Rota was born on April 27, 1932 in Vigevano, Italy. In 1956, he received his Ph.D. from Yale University, where his advisor was Jacob T. Schwartz. He is known for his outstanding work in such fields as functional analysis, combinatorial theory, and probability. He was a Hedrick lecturer (1967), AMS Colloquium Lecturer (1998), and also received the AMS Steele Prize (1988). Professor Rota died on April 18, 1999.

Rudin, Mary Ellen

Mary Ellen Rudin was born December 7, 1924 in Hillsboro, Texas. She obtained her Ph.D. in 1949 at the University of Texas under the supervision of R.L. Moore. Her work primarily centred upon set-theoretic topology. Besides publishing approximately seventy research papers on this subject, she was also involved in a large number of mathematical organizations, including the American Mathematical Society for which she was Vice President during 1980–1981.

Schwartz, Laurent-Moïse

Laurent-Moïse Schwartz was born March 5, 1915 in Paris, France. He obtained his doctorate from the Faculty of Science at Strasbourg in 1943. For his outstanding work on the theory of distribution, he became the first French mathematician to receive a Fields Medal in 1950. Not only did he contribute to many areas of mathematics, he was also actively involved in politics and found time to collect over 20,000 specimens of butterflies. Professor Schwartz died on July 4, 2002.

Serre, Jean-Pierre

Jean-Pierre Serre was born September 15, 1926 in Bages, France. He received his doctorate from the Sorbonne in 1951. He has been recognized for his extraordinary contributions to algebraic geometry, number theory and topology, by being awarded a Fields Medal in 1954 and the Abel Prize in 2003.

Shepp, Lawrence A.

Lawrence A. Shepp was born September 9, 1936. He obtained his Ph.D. from Princeton University in 1961, where his advisor was Willi Feller. He specializes in statistics and computational tomography. For his contributions, he has received many honours including the IEEE Distinguished Scientist Award in 1979.

Sinai, Yakov Grigorevich

Yakov Grigorevich Sinai was born September 21, 1935 in Moscow, Russia. He received his Ph.D. from Moscow State University in 1960, where his advisor was Andrey Kolmogorov. He is a pioneer in many fields of mathematics including dynamical systems, mathematical physics, and probability. For his outstanding contributions to both mathematics and physics, he has received many honours including the Boltzmann Medal in 1986, Dirac Medal in 1992, and the Wolf Prize in Mathematics in 1997.

Smale, Stephen

Stephen Smale was born July 15, 1930 in Flint, Michigan. He received his Ph.D. from the University of Michigan in 1957 under the supervision of Raoul Bott. In 1966, he received a Fields Medal for his work on differential geometry, especially on the generalised Poincaré conjecture. In 2007, he was awarded the Wolf Prize in mathematics for his groundbreaking contributions to differential topology, dynamical systems, and mathematical economics, amongst others.

Stanley, Richard Peter

Richard Peter Stanley was born June 23, 1944 in Larchmont, New York. He received his Ph.D. at Harvard University in 1971 under the supervision of Gian-Carlo Rota. He is recognized as a leading expert in combinatorics. For his outstanding contributions to both research and teaching, he has been given many honours including the 2001 Steele Prize and the 2003 Schock Prize.

Sternberg, Shlomo

Shlomo Sternberg was born on January 20, 1936, and earned his Ph.D. in 1957 from Johns Hopkins University, where his advisor was Aurel Wintner. He is recognized for his outstanding contributions in geometry, especially symplectic geometry. He was a Guggenheim Fellow at Harvard University in 1974, and is currently a professor at Harvard.

Taylor, Jean

Jean Taylor was born in San Mateo, California on September 17, 1944. She received her Ph.D. in 1973 from Princeton University under the supervision of Frederick Almgren, Jr. Her research has focused on problems related to soap bubbles and crystals, and how they evolve under various physical laws. She is a pioneer in the mathematics of minimal surfaces.

Thurston, William Paul

William Paul Thurston was born October 30, 1946 in Washington, D.C. He received his Ph.D. in 1972 from the University of California, Berkeley under the supervision of Morris Hirsch. Professor Thurston died on August 21, 2012.

Turing, Alan Mathison

Alan Mathison Turing was born June 23, 1912 in London, England. He obtained his Ph.D. in 1938 from Princeton University, where his advisor was Alonzo Church. Considered to be the father of modern computer science, he formalized the concept of the algorithm. During the Second World War, he contributed to breaking German naval codes. Professor Turing died on June 7, 1954.

Uhlenbeck, Karen Keskulla

Karen Uhlenbeck was born August 24, 1942 in Cleveland, Ohio. She received her Ph.D. from Brandeis University in 1968 and has since been a MacArthur Fellow in 1983, elected to the American Academy of Arts and Sciences (1985), the National Academy of Sciences (1986), and in 2000 she received a National Medal of Science for "special recognition by reason of [her] outstanding contributions to knowledge" in the area of mathematics. In 1990 she became only the second woman (after Emmy Noether in 1932) to give a Plenary Lecture at an International Congress of Mathematics.

Ulam, Stanisław Marcin

Stanisław Marcin Ulam was born April 3, 1909 in Lemberg, Poland. He received his Ph.D. in 1933 from the University of Lwów, where his advisor was Stefan Banach. He is known for solving the problem of how to initiate fusion in the hydrogen bomb as well as devising the `Monte-Carlo method', but he also contributed widely to set theory, ergodic theory, and group theory. Professor Ulam died on May 13, 1984.

Wahba, Grace

Grace Wahba was born August 3, 1934. She received her Ph.D. from Stanford University in 1966, where her advisor was Emanuel Parzen.

She is a pioneer in methods for smoothing noisy data, and her research interests include statistical model building, and numerical weather prediction.

Weil, André

André Weil was born May 6, 1906 in Paris, France. He received his doctorate from the University of Paris, supervised by Jacques Hadamard. For his fundamental work in number theory and algebraic geometry, he is recognized as one of the greatest mathematicians of the 20th century. He was also a founding member of the Bourbaki group. Professor Weil died on August 6, 1998.

Weinstein, Alan David

Alan David Weinstein was born June 17, 1943 in New York. He received his Ph.D. in 1967 from the University of California, Berkeley, where his advisor was Shiing-Shen Chern. He is best known for his contributions to symplectic geometry. Currently, he serves as the chair of the Department of Mathematics at Berkeley.

Wiles, Andrew John

Andrew John Wiles was born April 11, 1953 in Cambridge, England. He received his Ph.D. in 1979 from the University of Cambridge, where his advisor was John Coates. In 1995, he proved Fermat's Last Theorem. For this outstanding work, he was awarded the Schock Prize in 1995, and the Wolf Prize in 1996.

Bibliography

[ν] Original Material.

[1] Aczel, Amir D. *Descartes' Secret Notebook: A True Tale of Mathematics, Mysticism, and the Quest to Understand.* Broadway Books, New York, NY, 2006.

[2] Albers, Donald J. "Freeman Dyson: Mathematician, Physicist, and Writer", *The College Mathematics Journal* 25, no. 1 (January 1994): 2–21.

[3] —— "John Horton Conway—Talking a Good Game", *Math Horizons* Spring 1994, 6–8.

[4] Albers, Donald J. & Gerald L. Alexanderson. *Mathematical People: Profiles and Interviews*, 2nd Edition. A K Peters, Ltd., Natick, MA, 2008.

[5] Albers, Donald J., Gerald L. Alexanderson, & Constance Reid. *More Mathematical People*. Academic Press (Elsevier), Burlington, MA, 1994.

[6] Albers, Donald J., Gerald L. Alexanderson, & Richard K. Guy. "A Conversation with Richard K. Guy", *The College Mathematics Journal* 24, no. 2 (March 1993): 123–148.

[7] d'Alembert, Jean Le Rond. *Preliminary Discourse to the Encyclodpedia of Diderot*, translated by Richard N. Schwab. The University of Chicago Press, Chicago, IL, 1995.

[8] Alexander, J. W. "An Example of a Simply Connected Surface Bounding a Region Which is Not Simply Connected", *Proceedings of the National Academy of Sciences* 10, no. 1 (January 1924): 8–10.

[9] Aristotle. *Metaphysics, Book VIII, Chapter 3*. In *The Complete Works of Aristotle*, (Jonathan Barnes, ed.), Vol. 2. Princeton University Press, Princeton, NJ, 1985.

[10] Babai, László & Joel Spencer. "Paul Erdős (1913–1996)", *Notices of the American Mathematical Society* 45, no. 1 (January 1998): 64–73.

[11] Bacon, Roger. Opus Majus, Part 4, Distinctia Prima, cap. 1

[12] Barrow, Isaac. *The Usefulness of Mathematical Learning: Being Mathematical Lectures Read in the Publick Schools at the University of Cambridge*. Routledge, New York, NY, 1970.

[13] Barrow, John D. *Theories of Everything: the Quest for Ultimate Explanation*. Oxford University Press, New York, NY, 1991.

[14] Bell, Eric Temple. *Mathematics: Queen and Servant of Science*. McGraw–Hill Book Co., Inc., New York, NY, 1940.

[15] ——. *Gauss, the Prince of Mathematicians*, reprinted in *The World of Mathematics*, J. R. Newman (ed.), Simon and Schuster, New York, NY, 1956.

[16] ——. *The Queen of the Sciences*. The Williams & Wilkins Company, Baltimore, 1931.

[17] Bellman, Richard. *A Brief Introduction to Theta Functions*. Holt, Rinehart and Winston, New York, NY, 1961.

[18] ——. *Eye of the Hurricane: An Autobiography*. World Scientific Pub. Co. Inc., Hackensack, NJ, 1984.

[19] Birkhoff, George David. "Intuition, Reason and Faith in Science", *Science* 88, no. 2296 (December 1938): 601–609. Reprinted in *Musings of the Masters: An Anthology of Mathematical Reflections*, The Mathematical Association of America, 2004, pp. 98–114.

[20] Bôcher, Maxime. "The Fundamental Conceptions and Methods of Mathematics", *Bulletin of the American Mathematical Society* 11 (December 1904): 115–135.

[21] Bochner, Salomon. "Why Mathematics Grows", *Jour. Hist. Ideas* 26 (1965): 3–24. Reprinted in *Collected Papers of Salomon Bochner*. American Mathematical Society, 1992.

[22] Brewster, David. *Memoirs of the Life, Writings, and Discoveries of Sir Isaac Newton*. Thomas Constable and Co., Edinburgh, 1855.

[23] Campbell Lewis & William Garnett. *The life of James Clerk Maxwell: With selections from his correspondence and occasional writings*. Macmillan and Co., UK, 1884.

[24] Cayley, Arthur. "Section A. Mathematical and Physical Science. Inaugural Address by President, Prof. Arthur Cayley, M.A.". Reprinted in Norman Lockyer's *Nature: International Journal of Science*, v. 28. Nature Publishing Group, Macmillan Journals ltd., 1883.

[25] Chandrasekharan, K. *Hermann Weyl, 1885–1985: Centenary Lectures.* Springer Publishing Company, New York, NY, 1986.

[26] Courant, Richard. "Mathematics in the Modern World", *Scientific American* 211, no. 3 (September 1964): 41–49.

[27] Courant, Richard & David Hilbert. *Methods of Mathematical Physics*. Volume I. Interscience Publishers, New York, NY, 1953.

[28] Dantzig, Tobias. *Number: The Language of Science.* The Macmillan Company, New York, NY, 1954.

[29] Darboux, G. *La vie et l'oeuvre de Charles Hermite.* Revue du mois, 10 January 1906.

[30] Dehn, Max. "The Mentality of the Mathematician: A Characterization", translated by Abe Shenitzer, *Mathematical Intelligencer* 5, no. 2 (1983): 18–26.

[31] Descartes, René. *Discourse on Method and Related Writings*, translated by Desmond M. Clarke. Penguin Classics, 2000.

[32] ——. *Rules for the Direction of the Mind, Rule IX* Bobbs–Merrill Co., Indianapolis, Indiana, 2000.

[33] Dirac, Paul. "The Evolution of the Physicist's Picture of Nature", *Scientific American* 208, no. 5 (May 1963), 45–53.

[34] Dirichlet, J. P. G. L. *Werke, Bd. 2*. Chelsea Pub. Co., Bronx, NY, 1987.

[35] Dukas, H.& B. Hoffmann (eds.) *Albert Einstein, the Human Side: New glimpses from his archives.* Princeton University Press, Princeton, NJ, 1979.

[36] Dunnington, Guy Waldo, Jeremy Gray, and Fritz-Egbert Dohse. *Carl Friedrich Gauss: Titan of Science.* The Mathematical Association of America, Washington, DC, 2004.

[37] Duporcq, Ernest. *Compte rendu du deuxiéme congrès international des mathématiciens*. Gauthier–Villars, Paris, 1902.

[38] Dyson, Freeman J. "Birds and Frogs", *Notices of the American Mathematical Society* 56, no. 2 (February 2009): 212–223.

[39] ——. "Prof. Hermann Weyl, For.Mem.R.S.", *Nature* 177, no. 4506 (March 10, 1956): 457–458.

[40] Einstein, Albert & Paul Arthur Schilpp. *Albert Einstein: Philosopher–scientist*. Tudor Publishing Company, New York, NY, 1951.

[41] Eisenstein, Gotthold. *Mathematische Werke*, Preface by Gauss. Chelsea, New York, NY, 1975.

[42] Emmer, Michele. "Interview with Ennio De Giorgi", *Notices of the American Mathematical Society* 44, no. 9 (October 1997): 1097–1101.

[43] Euler, Leonhard. *Opera Omnia*, ser.1, vol.2.

[44] Fitzgerald, Michael & Ioan Mackenzie James. *The Mind of the Mathematician*. The John Hopkins University Press, Baltimore, MD, 2007.

[45] Fourier, Joseph. *The Analytical Theory of Heat*. Cambridge University Press, Cambridge, 1878.

[46] Galilei, Galileo. *Opere Il Saggiatore*.

[47] Galois, Évariste. *Manuscrits de Évariste Galois*. (Jules Tannery, ed.) Gauthier–Villars, Paris, 1908.

[48] Glaisher, J. W. L. "Presidential Address British Association for the Advancement of Science" (Section A), *Nature* 42, no. 1089 (September 1890): 464–468.

[49] Gödel, Kurt. "What is Cantor's Continuum Problem?", 1947, pp. 258–273 in *Philosophy of Mathematics: Selected Readings*, Second Edition, (Benacerraf, Paul, and Putnam, Hilary, eds.), Cambridge University Press, New York, NY, 1983.

[50] Grothendieck, Alexandre. Récoltes et Semailles: *Réflexions et témoignages sur un passé de mathématicien*, Université des Sci-

ences et Techniques du Languedoc, Montpellier, et Centre National de la Recherche Scientifique, 1986.

[51] Hadamard, Jacques. *Essay on the Psychology of Invention in the Mathematical Field.* Dover Publications, New York, NY, 1945.

[52] Halmos, Paul. *I Want To Be A Mathematician.* MAA, Springer–Verlag, New York, 1985.

[53] ——. "Mathematics as a Creative Art", *American Scientist* 56 (Winter 1968): 375–389.

[54] Hankel, Hermann. *Die Entwickelung der Mathematik in den letzten Jahrhunderten.* Tübingen, 1884.

[55] Hardy, G. H. *A Mathematician's Apology*. Cambridge University Press, Cambridge, 1992.

[56] ——. "Mathematical Proof", *Mind* 38 (1929): 1–12. Reprinted in *Musings of the Masters: An Anthology of Mathematical Reflections*. The Mathematical Association of America, 2004, pp. 48–63.

[57] ——. "On Prime Numbers", *Reports on the State of Science*, British Association, Manchester, 1915. Reprinted in Peter B. Borwein's *The Riemann Hypothesis*. Springer, 2008.

[58] Hilbert, David. Geometry and the Imagination.American Mathematical Society, Providence, RI, 1999.

[59] ——. "Logic and the Understanding of Nature" (Naturerkennen und Logik), *Naturwissenschaften* (1930): 959–963. Reprinted in *Musings of the Masters: An Anthology of Mathematical Reflections*, The Mathematical Association of America, 2004, pp. 118–127.

[60] ——. "Mathematical Problems" (Lecture delivered before the International Congress of Mathematicians at Paris in 1900), *Bulletin of the American Mathematical Society* 8, no. 10 (1902): 437–479, translated by Mary Winston Newson from the original, which appeared in the *Göttinger Nachrichten*, 1900, 253–297.

[61] ——. "über das Unendliche", *Mathematische Annalen* 95 (1926): 161–190. English translation by Stefan Bauer–Mengelberg, "On the infinite", in Jean van Heijenoort (ed.), *From Frege to Gödel:*

a source book in mathematical logic 1879–1931, Harvard University Press, MA, 1967, pp. 367–392.

[62] Hobson, Ernest William. "Presidential Address British Association for the Advancement of Science", *Nature* 84 (1910).

[63] Hodges, Andrew. *Alan Turing: the Enigma*. Simon and Schuster, 1983.

[64] Hoffman, Paul. *The Man Who Loved Only Numbers: The Story of Paul Erdős and the Search for Mathematical Truth*. Hyperion, London, 1999.

[65] Hölder, O. *Die Mathematische Methode*. Springer, Berlin, 1924.

[66] Hutton, Charles. *A Philosophical and Mathematical Dictionary, Vol. 1*. London, 1815.

[67] Jackson, Allyn. "*Comme Appelé du Néant*——As If Summoned from the Void: The Life of Alexandre Grothendieck", *Notices of the American Mathematical Society* 51, no. 10 (November 2004): 1196–1212.

[68] ——. "Interview with Louis Nirenberg", *Notices of the American Mathematical Society* 49, no. 4 (April 2002): 441–449.

[69] ——. "Interview with Shiing Shen Chern", *Notices of the American Mathematical Society* 45, no. 7 (August 1998): 860–865.

[70] Jourdain, Philip E. B. "The Nature of Mathematics", reprinted in *The World of Mathematics*, J. R. Newman (ed.), Simon and Schuster, New York, NY, 1956.

[71] Kac, Mark. *Enigmas of Chance*. Harper & Row, New York, NY, 1985.

[72] Kasner, Edward. "The Present Problems of Geometry", *Bulletin of the American Mathematical Society* 11, no. 6 (March 1905): 283–314.

[73] Kasner, Edward. & James Roy Newman. *Mathematics and the Imagination*. Courier Dover Publications, Mineola, NY, 2001.

[74] Keyser, Cassius Jackson. *Lectures on Science, Philosophy and Art*. Columbia University Press, New York, NY, 1908.

[75] ——. "Mathematical Emancipations", *Monist* 16 (1906).

[76] ——. "The Humanization of the Teaching of Mathematics", *Science, New Series* 35, no. 904 (April 26, 1912): 637–647.

[77] ——. "The Universe and Beyond", *Hibbert Journal 3* (1904–1905): 300–314.

[78] Khaldun, Ibn. *The Muqaddimah: An Introduction to History, vol. 3* translated by Franz Rosenthal. Bollingen, Princeton, NJ, 1969.

[79] Khayyam, Omar. *Treatise on Demonstration of Problems of Algebra*. 1070.

[80] Klein, Felix. " Aufgabe und Methode des mathematischen Unterrichts an den Universitäten", *Jahresbericht der Deutschen Mathematiker–Vereinigung* 7 (1899): 126–137.

[81] ——. *Development of Mathematics in the 19th Century*. Math. Sci. Press. Brookline, MA, 1979.

[82] ——. *Elementary Mathematics from an Advanced Standpoint: Arithmetic, Algebra, Analysis*, translated by E. R. Hedrick & C. A. Noble. Courier Dover Publications, Mineola, NY, 2004.

[83] Kline, Morris. *Mathematics in Western Culture*. Oxford University Press, New York, NY, 1964.

[84] Knopp, Konrad. "Mathematics as a Cultural Activity", translated by W. K. Bühler. *Mathematical Intelligencer* 7, no. 1 (1985): 7–14.

[85] Koenigsberger, Leo. *Carl Gustav Jacob Jacobi*. B. G. Teubner, Leipzig, 1904.

[86] Kronecker, Leopold. "Ueber den Zahlbegriff", *Journal für die reine und angewandte Mathematik* 101 (1887): 337–355.

[87] Krull, Wolfgang. "The Aesthetic Viewpoint in Mathematics", translated by Betty S. Waterhouse & William C. Waterhouse. *Mathematical Intelligencer* 9, no. 1(1987): 48–52.

[88] Lakatos, Imre. *Mathematics, Science, and Epistemology*. Cambridge University Press, Cambridge, 1978.

[89] Landau, Susan. "In Her Own Words: Six Mathematicians Comment on Their Lives and Careers", *Notices of the American Mathematical Society* 38, no. 7 (September 1991): 702–706.

[90] Lang, Serge. *The Beauty of Doing Mathematics: Three Public Dialogues*. Springer–Verlag, New York, 1985.

[91] Langlands, R. P. "Harish–Chandra. 11 October 1923–16 October 1983", *Biographical Memoirs of Fellows of the Royal Society* 31 (November 1985): 198–225.

[92] Laplace, Pierre–Simon. T*héorie Analytique des Probabilités, Second Partie.* MME V^{E} Courcier, Paris, 1812.

[93] Laugwitz, Detlef. *Bernhard Riemann 1826–1866: Turning Points in the Conception of Mathematics*, translated by Abe Shenitzer. Birkhäuser, Boston, MA, 1999.

[94] Leibniz, Gottfried Wilhelm. *The Early Mathematical Manuscripts of Leibniz*, translated by James Mark Child. The Open Court Publishing Company, Chicago, Illinois, 1920.

[95] ——. *New Essays Concerning Human Understanding, IV, XII* translated by Alfred Gideon Langley. The Open Court Publishing Company, Chicago, Illinois, 1916.

[96] Lichnerowicz, André. "The Community of Scholars" (La communauté des savants), *Enseignement Math.* 1, sér 2 (1955): 30–43. Translated from the French in *Musings of the Masters: An Anthology of Mathematical Reflections*, The Mathematical Association of America, 2004, pp.185–198 .

[97] Littlewood, John Edensor. *Littlewood's Miscellany*. Revised edition by Béla Bollobás, Cambridge University Press, Cambridge, 1986.

[98] Mandelbrot, Benoit B. and Richard L. Hudson. *The Misbehavior of Markets: A Fractal View of Risk, Ruin, and Reward*, Basic Books, New York, NY, 2006.

[99] Mashaal, Maurice. *Bourbaki: A Secret Society of Mathematicians*, translated by Anna Pierrehumbert. American Mathematical Society, Providence, RI, 2006.

[100] Maxwell, James Clerk. *The Scientific Papers of James Clerk Maxwell*, Vol. 2. Courier Dover Publications, Mineola, NY, 2003.

[101] Mazur, Barry. "Number Theory as Gadfly", *American Mathematical Monthly* **98**, no. 7 (Aug.–Sep. 1991): 593–610.

[102] Monastyrsky, Michael. *Riemann, Topology, and Physics*, Birkhäuser, Boston, MA, 1999.

[103] Mordell, Louis Joel. *Three Lectures on Fermat's Last Theorem*, The University Press, Cambridge, 1921.

[104] Morse, Marston. "Mathematics and the Arts", *Bulletin of the Atomic Scientists* 15, no. 2 (1959): 55–59. Reprinted in *Musings of the Masters: An Anthology of Mathematical Reflections*, The Mathematical Association of America, 2004, pp. 81–94.

[105] Nemerov, Howard. *The Western Approaches: Poems, 1973–75*, University of Chicago Press, Chicago, 1975.

[106] Neumann, John von. "The Mathematician", in *John von Neumann, Collected Works*, Vol I, Pergamon, New York, NY, 1961, pp. 1–9. Reprinted in *Musings of the Masters: An Anthology of Mathematical Reflections*, The Mathematical Association of America, 2004, pp. 172–184.

[107] Neumann, John von, F. Bródy, and Tibor Vámos. *The Neumann Compendium*. World Scientific, London, 1995.

[108] Newman, James Roy (ed.) *The World of Mathematics*, Simon and Schuster, New York, NY, 1956.

[109] Newman, M. H. A. "What is Mathematics? New Answers to an Old Question", Presidential Address to the Mathematical Association, April 1959, *Mathematical Gazette* 43, no. 345 (October 1959): 161–171.

[110] O'Connor, J. J. & E. F. Robertson "Charles Louis Fefferman" http://www–groups.dcs.st–and.ac.uk/~history/Biographies/Fefferman.html. JOC/EFR, April 1998.

[111] Olson, Richard. *Scottish Philosophy and British Physics, 1750–1880*. Princeton University Press, 1975.

[112] Papert, Seymour A. *Mindstorms: Children, Computers, and Powerful Ideas*. Basic Books, New York, NY, 1993.

[113] Parry, Bill. "Two Poems by Bill Parry (1934–2006)", *Notices of the American Mathematical Society* 54, no. 3 (March 2007): 386–387.

[114] Pascal, Blaise. *Pensées*, translated by Alban J. Krailsheimer. Penguin Classics, 1995.

[115] Penrose, Roger. *The Emperor's New Mind: Concerning Computers, Minds and the Laws of Physics*. Oxford University Press, Oxford, 1999. (Orig. pub. 1989.)

[116] Peirce, Benjamin. "Linear Associative Algebra", *American Journal of Mathematics* 4, no. 1 (1881): 97–229.

[117] Peirce, Charles Sanders & Morris R. Cohen. *Chance, Love, and Logic: Philosophical Essays*, Routledge, London, 2000.

[118] Peirce, Charles Sanders. "The Essence of Mathematics", reprinted in *The World of Mathematics*, J. R. Newman (ed.), Simon and Schuster, New York, NY, 1956.

[119] Pétard, H. "A Brief Dictionary of Phrases Used in Mathematical Writing", *American Mathematical Monthly* 73, no. 2 (February 1966): 196–197.

[120] Phillips, Anthony. "Chih–Han Sah (1934–1997)", *Notices of the American Mathematical Society* 45, no. 1 (January 1998): 79–82.

[121] Pickover, Clifford A. *Computers and the Imagination*. St. Martin's Press, New York, NY, 1991.

[122] Plutarch. *Plutarch's Lives, Vol. 2, Marcellus* translated by John Langhorne and William Langhorne. London, 1798.

[123] Poincaré, Henri. "L'intuition et la logique en mathématiques", translated by G. B. Halsted in *The Foundations of Science*, The Science Press, New York, 1913. Also see the translation "Mathematical Creation" in *Mathematics in the Modern World*, pp. 14–17, and in *Scientific American*, August 1948.

[124] ——. *Science and Method*, translated by Francis Maitland. Cosimo, Inc, New York, NY, 2007.

[125] ——. *Value of Science*, translated by George Bruce Halsted. Cosimo Inc., New York, NY, 2007.

[126] Pólya, George. *How to Solve It*. Princeton University Press, Princeton, NJ, 1945.

[127] ——. *Mathematical discovery: On understanding, learning and teaching problem solving*, Vol. 2, Wiley, New York, NY, 1965.

[128] ——. *Mathematics and Plausible Reasoning*. Princeton University Press, Princeton, NJ, 1954.

[129] ——. "On Plausible Reasoning", *Proceedings of the International Congress of Mathematicians 1950, Vol. 1*, American Mathematical Society, Providence, RI, 1952.

[130] ——. "Two incidents", *Scientists at Work: Festschrift in Honour of Herman Wold*, ed. T. Dalenius, G. Karlsson, and S. Malmquist; Uppsala, Sweden: Almquist & Wiksells Boktryckeri AB, 1970, pp. 165–68.

[131] Pringsheim, Alfred. "über Wert und angeblichen Unwert der Mathematik", *Jahresbericht der Deutschen Mathematiker–Vereinigung* 13 (1904): 357–382.

[132] Raussen, Martin & Christian Skau. "Interview with Jean–Pierre Serre", *Notices of the American Mathematical Society* 51, no. 2 (February 2004): 210–214.

[133] ——. "Interview with Michael Atiyah and Isadore Singer", *Notices of the American Mathematical Society* 52, no. 2 (February 2005): 223–231.

[134] ——. "Interview with Peter D. Lax", *Notices of the American Mathematical Society* 53, no. 2 (February 2006): 223–229.

[135] Reid, Constance. "Being Julia Robinson's Sister", *Notices of the American Mathematical Society* 43, no. 12 (December 1996): 1486–1492.

[136] Rosanes, Jakob. "Charakteristische Züge in der Entwicklung der Mathematik des 19. Jahrhunderts", *Jahresbericht der Deutschen Mathematiker–Vereinigung* 13 (1904): 17–30.

[137] Rota, Gian–Carlo. *Indiscrete Thoughts*. Birkhäuser, Boston MA, 1997.

[138] Russell, Bertrand. *Mysticism and Logic and Other Essays*. Longmans Green and Co., London, 1918.

[139] ——. *Portraits From Memory, and Other Essays*, Simon and Schuster, New York, NY, 1956.

[140] ——. "Recent Work on the Principles of Mathematics". *International Monthly* 4 (1901). Reprinted in *Bertrand Russell: His Works*. Vol. 3: Toward the "Principles of Mathematics", 1900–02. Routledge, New York, NY, 1994.

[141] Safra, Edmond J. *Einstein's 1912 Manuscript on the Special Theory of Relativity*. George Braziller, New York, 2004.

[142] Schubert, Hermann. *Mathematical Essays and Recreations*, The Open Court Publishing Company, Chicago, Illinois, 1898.

[143] Schwartz, Laurent. *A Mathematician Grappling with His Century*, translated from the French by Leila Schneps. Birkhäuser Verlag, Basel, Switzerland, 2001.

[144] Smale, Stephen, Felipe Cucker, & Roderick Wong. *The Collected Papers of Stephen Smale*, World Scientific, 2000.

[145] Smith, David Eugene. *The Teaching of Geometry*, Ginn and Company, 1911.

[146] Smith, Henry J. S. "Presidential Address British Association for the Advancement of Science, Section A", *Nature* 8 (1873).

[147] Stanford University News Service. "David Donoho named MacArthur Fellow". 18 June 1991. Web. 31 August 2009. http://news–service.stanford.edu/pr/91/910618Arc1292.html

[148] Stevens, Wallace. *The Collected Poems of Wallace Stevens*.Vintage Books, 1990.

[149] Stoicheff, B. P. *Gerhard Herzberg: an illustrious life in science*. NRC Press, Ottawa, 2002.

[150] Sylvester, James Joseph. A Probationary Lecture on Geometry, Delivered Before the Gresham Committee and the Members of the Common Council of the City of London, 4 December, 1854; *The Collected Mathematical Papers of James Joseph Sylvester, Vol. 4*, University Press, Cambridge, 1908, 2–9.

[151] ——. Baltimore Address on Commemoration Day at John Hopkins University, 22 February, 1877; *The Collected Mathematical Papers of James Joseph Sylvester, Vol. 3*, University Press, Cambridge, 1908, 72–87.

[152] ——. On Newton's Rule for the Discovery of Imaginary Roots of Equations, Proceedings of the Royal Society of London, XIV, 1865, pp. 268–270; *The Collected Mathematical Papers of James Joseph Sylvester, Vol. 4*, University Press, Cambridge, 1908, 493–494.

[153] ——. Presidential Address to Section 'A' of the British Association, Exeter British Association Report, 1869, pp. 1–6; *The Collected Mathematical Papers of James Joseph Sylvester, Vol. 4*, University Press, Cambridge, 1908, 650–661.

[154] Szymborska, Wisława. *Poems, new and collected: translated from the Polish by Stanisław Barańczak and Clare Cavanagh.* Harcourt Inc, New York, NY, 1998.

[155] Tait, P.G. "Section A. Mathematical and Physical Science. Opening Address by the President, Prof. P.G. Tait, M.A.". Reprinted in Norman Lockyer's *Nature: International Journal of Science*, Nature Publishing Group, Macmillan Journals ltd., 1871.

[156] Thompson, Silvanus P. *The Life of William Thomson, Baron Kelvin of Largs*. Macmillan, 1910.

[157] Turán, Paul. "The Work of Alfréd Rényi", *Matematikai Lapok* 21 (1970): 199–210.

[158] Turing, Alan. *The Essential Turing: Seminal Writings in Computing, Logic, Philosophy, Artificial Intelligence*. Oxford University Press, Oxford, 2004.

[159] Uhlenbeck, Karen. "Interview with Tai–Ping Lee and Lan Hsuan Huang", Taipei, December 2005. Web. 31 August, 2009. http://rene.ma.utexas.edu/users/uhlen/vita/interview05.pdf

[160] Ulam, Stanisław. *Adventures of a Mathematician*. University of California Press, Berkeley, CA, 1991.

[161] Venn, John. *Symbolic Logic*. Macmillan and Co., London, 1881. Reprinted by American Mathematical Society, New York, 1971.

[162] Weber, H. "Obituary for Leopold Kronecker", *Jahresberichte der Deutschen Mathematiker–Vereinigung* 2 (1891/92): 5–31.

[163] Weil, André. "History of Mathematics: Why and How", *Proc. International Congress of Mathematicians, Helsinki, 1978*, Vol.

1, Academia Scientiarum Fennica, Helsinki, 1980, pp. 227–236. Reprinted in *Musings of the Masters: An Anthology of Mathematical Reflections*, The Mathematical Association of America, 2004, pp. 202–213.

[164] Weil, André. "Mathematical Teaching in Universities", *American Mathematical Monthly* 61, no. 1 (January 1954): 34–36.

[165] ——. *The Apprenticeship of a Mathematician*. Birkhäuser, Boston, MA, 1992.

[166] ——. "The Future of Mathematics", *American Mathematical Monthly* 57, no. 5 (May 1950): 295–306.

[167] Weyl, Hermann. *The Classical Groups: Their Invariants and Representations*. Princeton University Press, Princeton, NJ, 1946.

[168] ——. "The Unity of Knowledge," *H. Weyl, Gesammelte Abhandlungen*, Vol. 4, Springer Publishing Company, Berlin–Heidelberg, 1968, pp. 623–630. Reprinted in *Musings of the Masters: An Anthology of Mathematical Reflections*, The Mathematical Association of America, 2004, pp. 81–94.

[169] ——. "Part II. Topology and Abstract Algebra as Two Roads of Mathematical Comprehension", translated by Abe Shenitzer in *The American Mathematical Monthly* 102, no. 7 (August–September 1995): 646–651.

[170] Whitehead, Alfred North. *Adventures of Ideas*. The Macmillan Company, New York, NY, 1933.

[171] Whitehead, Alfred North & Donald W. Sherburne. *A Key to Whitehead's Process and Reality*. University of Chicago Press, Chicago, IL, 1981.

[172] ——. *An Introduction to Mathematics*. Oxford University Press, New York, NY, 1959.

[173] ——. "Inaugural Lecture, Oxford, 1885", *Nature* 33 (1885).

[174] ——. *Science and the Modern World: Lowell Lectures, 1925*. New American Library, New York, NY, 1956.

[175] Wiener, Norbert. *Ex–Prodigy: My Childhood and Youth*. Simon and Schuster, New York, NY, 1953. http://www.pbs.org/wgbh/nova/transcripts/2414proof.html

[176] Wiles, Andrew. NOVA Transcript of "The Proof".

[177] Wolffsohn, Lily. *Sónya Kovalévsky: Her Recollections of Childhood*, translated by Isabel F. Hapgood, Annie M. C. Bayley. The Century Co., 1895.

Index

Credits

Page 2. Martin Raussen and Christian Skau, "Interview with Michael Atiyah and Isadore Singer" *European Mathematical Society Newsletter* 2004 (24–30).

Page 4. "Pi" from POEMS, NEW AND COLLECTED: 1957–1997 by Wislawa Szymborska, English translation by Stanislaw Baranczak and Clare Cavanagh copyright © 1998 by Houghton Mifflin Harcourt Publishing Company, reprinted by permission of the publisher.

Page 16. "Figures of Thought" from *The Western Approaches: Poems: 1973–75*, by Howard Nemerov, Chicago Press, 1975, reprinted with permission of Alexander Nemerov.

Page 18. A Poem by Shiing Shen Chern reprinted with permission of T.Y. Lam.

Page 23. Richard Courant and Herbert Robbins; *What is Mathematics?: An elementary Approach to Ideas and Methods*, Oxford University Press, 1941. By permission of Oxford University Press, USA.

Page 33. Dyson, Freeman J. "Birds and Frogs," *Notices of the American Mathematical Society 56*, no. 2 (2009) pp. 212.

Page 37. Einstein, Albert; *Albert Einstein, The Human Side* © 1979 by the Estate of Albert Einstein, Princeton University Press. Reprinted by permission of Princeton University Press.

Page 41. Hoffman, Paul *The Man Who Loved Only Numbers*, Hyperion, New York, 1998

Page 62. G.H. Hardy, Foreword by C.P. Snow, *A Mathematician's Apology*, Cambridge University Press, 1940. Reprinted with permission of Cambridge University Press.

Page 86. Martin Raussen and Christian Skau, "Interview with Peter D. Lax" *European Mathematical Society Newsletter* 2005 (24–31).

Page 101. "The Mathematician" from *The Works of Mind*, by Robert B. Heywood, University of Chicago Press, Chicago, 1966, pp. 180–96.

Page 103. Jackson, Allyn, "Interview with Louis Nirenberg" Notices of the American Mathematical Society 45, no. 1 2002 (449)

Page 112. Polya, George, "Mathematics and Plausible Reasoning" in *Proceedings of the International Congress of Mathematicians* 1950, Vol. 1, American Mathematical Society, Providence, RI, 1952.

Page 116. Reid, Constance, "Being Julia Robinson's Sister" *Notices of the American Mathematical Society* 43, no. 12 1996 (1492). By permission of Neil D. Reid.

Page 125. Martin Raussen and Christian Skau, "Interview with Jean-Pierre Serre" *European Mathematical Society Newsletter* 2003 (18–20).

Page 141. Venn, John *Symbolic Logic*, Macmillan and Co., London, 1881. Reprinted by American Mathematical Society, New York, 1971.

Items on pages 8–9, 11–12, 19, 20–21, 25–26, 50–52, 55, 71–72, 84, 94, 112–113, 115–116, 123, 125, and 131–132 are taken from *More Mathematical People*, Gerald L. Alexanderson, Donald J. Albers, and Constance Reid, Harcourt, Brace, Jovanovich, 1990. Reprinted with permission from the authors.

Items on pages 30–31 and 79 are taken from *Mathematical People*, Gerald L. Alexanderson and Donald J. Albers, Birkhauser Boston, Inc., 1985. Reprinted with permission from the authors.

Despite every effort to contact copyright holders and obtain permission prior to publication, in some cases this has not been possible. If notified, we will undertake to rectify any errors or omissions at the earliest opportunity.

About the Editors

Peter Borwein

Peter Borwein is the founding Project Leader and currently an Executive Co-Director of The IRMACS Centre. He is a Burnaby Mountain Chair at Simon Fraser University and has been a professor in the mathematics department since 1993 when he moved from Dalhousie University. He is also an adjunct professor in computing sicence. His research interests span various aspects of mathematics and computer science, health and criminology modelling and visualization.

Peter Liljedahl

Peter Liljedahl is an associate professor of mathematics education in the Faculty of Education, an associate member in the department of mathematics, and co-director of the David Wheeler Institute for Research in Mathematics Education at Simon Fraser University in Vancouver, Canada. His research interests are creativity, insight, and discovery in mathematics teaching and learning; the role of the affective domain on the teaching and learning of mathematics; the professional growth of mathematics teachers; mathematical problem solving; and numeracy.

Helen Zhai

Helen Zhai graduated with a BSc in mathematics and BEd from Simon Fraser University. She has received undergraduate NSERC grants, one of which initiated her collaboration with Peter Borwein and Peter Liljedahl in their work on creativity in mathematics teaching and learning.